Design of

CURVED MEMBERS

for Machines

Design of CURVED MEMBERS for Machines

ALEXANDER BLAKE

Technical Specialist, Aerojet General Corporation
Sacramento, California

ROBERT E. KREIGER PUBLISHING COMPANY
HUNTINGTON, NEW YORK
1979

Original Edition 1966
Reprint 1979 with corrections

Printed and Published by
ROBERT E. KRIEGER PUBLISHING COMPANY, INC.
645 NEW YORK AVENUE
HUNTINGTON, NEW YORK 11743

Copyright © 1966 by
INDUSTRIAL PRESS, INC.
Transferred to Alexander Blake 1978
Reprinted by Arrangement

*All rights reserved. No reproduction in any form of this
book, in whole or in part (except for brief quotation in
critical articles or reviews), may be made without written
authorization from the publisher.*

Printed in the United States of America

Library of Congress Cataloging in Publication Data

Blake, Alexander.
   Design of curved members for machines.

   Reprint of the edition published by Industrial Press, New
York, with corrections.
   Bibliography: p.
   Includes index.
   1. Machinery—Design.   2. Machine parts.   I. Title.
[TJ233.B6 1979]   621.8'15   79-12202
ISBN 0-88275-970-1

621.815
B636d

To Iris and Susan

Preface

The overall concept of this book is to bring to focus the important elastic equations for the design of a variety of curved members involved in construction of machines and supporting hardware. The text contains numerous ready-to-use formulas, charts and numerical examples applicable to circular rings, arched cantilevers, flat mechanical springs, curved beams, hooks, machine frames, circular arches, support brackets and similar components under various conditions of loading and support. The book brings together many formulas not found elsewhere or lost in the flood of current technical literature. The design topics and data presented should find application in all areas of industry wherever strength and rigidity characteristics of non-straight elastic members are of importance. Although the designs based on a linear theory of elasticity err generally on the conservative side, various practical considerations and safety requirements of modern machinery make the use of elastic analysis mandatory.

The treatment of the subject matter is generally tailored to the needs of the practicing engineer and the theory is discussed only if it illuminates the derivation of working equations. The proofs of the theorems used are considered not essential to follow the mathematical techniques involved and are, therefore, omitted to limit undue repetition of the theoretical details adequately covered elsewhere. However, the analysis outlined often includes those steps in derivations, considered essential to grasping the physical and mathematical concepts of design, which are sometimes labelled "obvious" and thereby excluded from otherwise useful engineering literature.

The derivations of stress and deflection formulas for curved elastic members presented in this book are based essentially on the Winkler-

Bach theory, equations of the elastic line for curved bars and the classical theorems of Castigliano. Particular emphasis is placed on the application of the theorems of Castigliano which are powerful tools in solving statically determinate and indeterminate problems. The formulas and numerical data discussed together with the principle of superposition, can be applied to great many practical design situations. Furthermore, the derivation techniques can be extended to a variety of machine design problems involving the geometry and loading conditions not treated specifically in this volume.

Although this book is written primarily for the practicing engineer and designer it is hoped that the material presented will prove useful to members of the teaching profession, concerned with machine design and structural analysis, and to the recently graduated engineers making rather drastic transition from the academic environment to the industrial business routine.

Over the past several years many individuals and organizations in this country and England, engaged in design and research, influenced the development of this text by means of personal contact or written communications. Furthermore, a significant number of formulas and charts, featured in this book, have been published from time to time in Machine Design, Product Engineering and other magazines or manuals. Such contributions are hereby gratefully acknowledged.

The author wishes to express his thanks for the help received from his friends and associates at the Aerojet-General Corporation during various stages of the preparation of the manuscript. Mr. John G. Schumacher contributed to the development of material for sections on rings, arcuate beams and arched cantilevers and looked over the complete draft. Dr. Wilhelm Riedel and Dr. Karl Morghen read the final manuscript and offered useful comments. Mr. William A. Lester contributed isometric illustrations which added to the clarity of presentation. Mr. Sigmunt R. Roland provided assistance in the final preparation of the design charts and diagrams for various sections of the book. Mrs. Sharon L. Adamski typed the preliminary material and the complete manuscript.

Special thanks are due to Professor Kenneth E. Lofgren for his review of the sample material and the final manuscript as well as his direct contribution to the book by providing the analysis of curved beams with varying cross-sections. Last but not least, the author is indebted to Mr. Holbrook L. Horton and Mr. Paul B. Schubert of the Industrial Press for their encouragement and editorial help during all phases of getting this book into print.

Finally, although in developing and writing this book no effort has been spared to assure the correctness of formulas, tables and numerical work, it is perhaps, too much to hope that all errors have been avoided. The author therefore welcomes the calling to his attention of any of the shortcomings and errors which may be discovered.

ALEXANDER BLAKE

Rancho Cordova, California
October, 1965

Contents

List of Tables

CHAPTER ONE

General
Considerations

Introduction

The demand for modern machinery and elaborate supporting equipment has been increasing constantly with the accelerated pace of postwar technology. Answers to many theoretical questions, once of academic interest only, find new areas of application. The older cut-and-try methods of machine design become largely uneconomical while the practicing engineer is frequently required to make on-the-spot decisions. Since to design well is to make sound decisions a comprehensive knowledge of the mathematical techniques, materials science and loading conditions should be combined for best results. Although an immense volume of technical literature is available the elastic analysis of even simple load carrying members is often tedious. Since many machine systems consist of complex arrangement of straight and curved members the detailed stress analyses of modern machinery becomes difficult and time consuming even for elementary conditions of loading and support. Especially wide use of curved machine and structural members of almost infinite variety in present-day industries demands that the design theory and the important design equations in this field of knowledge be brought to focus.

Numerous machine elements and structural components, whose center lines are plane curves, may be classified under the general heading of curved members. Furthermore, in many practical situations, the majority of curved members possess center lines which can be approximated by circular curvature without appreciable error. Such members include closed and split rings, chain links, hooks, springs, arched cantilevers, curved beams, arches and similar machine elements. These members may be loaded tranversely or in their planes of curvature, and supported in a statically determinate or indeterminate manner.

The design analysis of curved members for machines outlined in this book is based on the theory of strength of materials. By the definition, elastic behavior requires that the extensions are recoverable on removal of the loads. Since most common engineering materials display linear elastic relationship the concept of linear theory of elasticity applies to the majority of the theoretical considerations presented. Although the designs based on a linear theory of elasticity err generally on the conservative side, various practical considerations and safety requirements of modern machinery make the use of the elastic analysis mandatory.

The development of almost all machines known to date has been influenced by past design experience derived largely from empirical relationships. Today, the extent to which the engineers have to rely on empirical methods, has been reduced considerably because of the progress in analytical techniques. Selection of suitable mathematical models leads to design formulas describing geometrical and physical relations of a particular load carrying member. However it should be noted that the mathematical approach has certain limitations with regard to physical significance of the numerical data derived from the theoretical equations.

The structural integrity of a machine part is defined largely by the magnitudes of stresses and deformation under specific working conditions. The strength of a material is determined by intensities of force at a given cross-section and measured usually in pounds per square inch. The load on a structure represents an external force acting upon a physical boundary. The magnitudes of the stresses influence the geometry, size and material of the structure member under consideration. Since an infinitely rigid material does not exist, machine elements and support members experience small but finite deflections under working loads. Whereas stress is an artificial concept which cannot be perceived or measured directly, strain can be recorded in many ways and then used in deriving the corresponding stress.

Arcuate Members for Machines

As machine components split rings sometimes known as arched cantilevers or arcuate beams may be subjected to forces that are in plane, out-of-plane or both. In the analysis of such members one end is usually considered fixed while the other is assumed to be displaced under the application of forces or moments.

When the forces acting on a curved bar do not lie in the plane of the curvature it is necessary to consider the deflection in two perpendicular planes in addition to the twist of the bar cross-sections. The analysis of out-of-plane bending of these elements can be applied to the design of single-turn helical springs and various retaining brackets. However the most important practical problem to which this analysis is applicable concerns the design of frames or piping subjected to temperature changes and movement of support reactions. The basic problem of the determination of stress distribution on various elements of such structures becomes statically indeterminate, the degree of redundancy being dependent on the manner of support. When the relation between the applied forces and the corresponding displacements of the free end is established the magnitude and the direction of moment or thrust can be found. In the particular case of transverse loading the torsional as well as bending stresses should be evaluated.

Curved-end cantilevers and complex-shape springs may be defined as machine and structural elements consisting of straight and curved portions, which find application as structural support brackets, machine members, fasteners and mechanical springs. The variety of shapes is truly enormous because of the complexity of modern machinery. However, using the method of superposition and the design equations developed for a few basic geometries, many practical stress problems in statically determinate as well as complex redundant structures can be solved within the acceptable limits of accuracy. What constitutes the acceptable limit in this particular case is really a matter of sound engineering judgement compatible with the specific operating requirements of a machine part as well as the knowledge of material properties. To make a better educated guess the mechanics of deformation and stress behavior of a structural member should be analyzed in some detail. Probably one of the most direct approaches to gaining appreciation of the basic design philosophy is through the development of the design equations and putting them to work for you.

Complex Springs

Determination of bending stresses in curved-end cantilevers and complex springs follows the elementary rules of strength of materials. The bending moment is simply the load times the shortest distance between the point of application of the load and the point at which bending is considered. The corresponding bending stress can be then calculated from the conventional straight beam expression utilizing the familiar concept of section modulus. Such calculations are usually straightforward when beam dimensions and loading are known. However, in dealing with complex curved members it is first advisable to investigate the bending moment distribution along the contour of the beam in order to locate the regions of zero bending slope or zero bending moment so that the structure in question could be subdivided into simpler components for the purpose of expediting the analysis. The geometry of the individual components can then be related to the stresses and strains on the assumption that the theory of pure flexure and Hooke's law apply. The theoretical approach to the solution of these problems is satisfactory for many applications to electro-mechanical and high-precision mechanical devices which are invariably required to work under elastic conditions.

The analysis usually called for in estimating deflections and stresses of complex load carrying members such as for instance, flat springs, is frequently avoided by having a sample built in the shop and then subjecting it to tests. Aside from the obvious time and cost considerations (the first or even second design seldom suffice), this approach results in the following disadvantages:

(1) It does not predict the maximum stress — a requirement if a factor of safety is to be known or where fatigue is involved.

(2) The sample may differ subtly from those made later by mass production techniques — often leading to failures or undesirable performance. Where the springs and similar members consist of circular elements, straight elements connected by small radii and those with both circular and curved portions the design theory presented here is based on the simple analytical approach for finding deflections and maximum stresses. Various key steps outlined in deriving the design equations are intended as a guide for the designer in the development of formulas for a variety of shapes not analyzed in this book. The equations are based on the assumption that the spring-like members are sufficiently stable laterally so that only bending in the plane of curvature requires attention.

Closed Rings

A type of curved member occurring very frequently in machine design is an elastic ring which can be subjected to in-plane or out-of-plane loading. A knowledge of the strength and deformation characteristics of elastic rings is essential to mechanical and structural design in many branches of engineering. While a few classical formulas for stresses and deflections in circular rings are found in most textbooks on strength of materials it is not easy to find the complete set of ready-to-use design equations for in-plane and out-of-plane loading conditions. The most elementary applications are those where elastic rings perform individually as free members in contrast to shell-supported rings where tangential shear forces distributed around the ring periphery are required for maintaining the static equilibrium. The theory and design equations of individual continuous rings can be combined by super-position to cover a variety of conditions of loading and support. An important area of machine design pertains to out-of-plane deformation of a circular elastic ring of uniform and compact cross-section encountered in the design of a frame supported on trunnions. The practical problems in which such curved members are subjected to the combined effect of bending and torsion, include hoist mechanisms, support skirts, gyroscope gimbals, foundations of storage tanks and similar supporting hardware. The so called toroidal deformation or inversion of a circular ring can be caused by twisting couples uniformly distributed along its center line. Typical examples of this type of loading condition are found in pipe flanges or retaining rings of commutators in electric motors.

Curved Beams and Hooks

In many design situations the cross-sectional dimensions of a curved member are small compared with the mean diameter and consequently the neutral axis coincides with the central axis. However in designing crane hooks, chain links, eye-shaped ends of bars, heavy gears, thick proving rings, machine frames and other machine elements characterized by sharp curvature the calculated stresses and deflections should be corrected for the displacement of the neutral axis in relation to the centroid of the section. Furthermore in such cases the effects of shear and direct stresses should be considered in addition to the customary analysis of bending. A very important manufacturing area is concerned with the rigidity of machine frames where elastic deflection constitutes a recognized design criterion. Such frames are shown to be analyzed as curved beams of variable cross-section.

Circular Frames

To facilitate the application of circular arches as machine components, some simplified equations and charts are given for determining moments and deflections according to the following conditions of support: free, pin-jointed and fixed. In more frequently encountered machine design applications the arch carries a concentrated central load in the plane of curvature. The cross-sectional dimensions of the arcuate beams, arched cantilevers or arches are normally assumed to be uniform and small in relation to their radii of curvature. Thus the neutral axis coincides with the central axis. The material is considered to be elastic and the deformation is caused mainly by bending. Consequently the effect of normal forces and the transverse shear on the stress and deformation characteristics of such curved members can be neglected without undue loss of accuracy. The exception to this practice concerns flat hingeless arches where the effect of normal forces becomes appreciable. This effect is known sometimes as rib shortening. The general rule of thumb is that when the arch rise to length ratio is greater than 0.2 the effect of axial deformation is normally ignored.

In most cases the supports are assumed to be unyielding and no elastic strain is great enough to alter significantly the original circular curvature. The assumption of unyielding supports however can be sometimes waived when the displacement of a support is specified on account of thermal effects or interface mislocation under assembly conditions.

In civil engineering practice the analysis of arched frames is usually based on parabolic rather than circular curvature. This means that the coordinates of the axis of a curved member are defined by the quadratic equation involving rise and span of the arch. For a relatively flat arch circular and parabolic curvatures are approximately the same and the solutions evolved for circular arches should apply reasonably well to parabolic arches.

CHAPTER TWO

Early Developments of Design Theory For Curved Members

Introduction

The types of curved members usually encountered in mechanical and structural engineering were briefly discussed in Chapter 1. Because of various theoretical and practical considerations the center lines of such members are often designed and made to circular curvature. This chapter is the outcome of a survey of the literature on the subject of analytical and experimental work related to the design of curved members the center lines of which can be approximated by circular arcs.

With the inevitable trend in industry toward greater complexity of machinery many theoretical questions involving curved structural members require more precise answers. To promote appreciation of the basic philosophy underlying design theory of such members it may be of interest to quote a few facts from early history leading to the development of the present level of state-of-the-art in this important area of strength of materials.

Most load carrying members such as lifting gear components, hooks, chain links and rings have always had varied industrial applications. Yet experimental verification of the relevant theories has received rather scant attention from early investigators in the field of elasticity and strength of materials. For instance before the year of 1900 hardly any experiments were made for the purpose of testing a theory and the formulas for calculating the ultimate strength and rigidity of curved members were purely empirical. While laboratory investigations in those days were naturally restricted because of the lack of suitable apparatus and reliable experimental techniques the development of the basic design theory was marked with considerable success. For example, Winkler and Résal developed the approximate theory for the distribution of normal stresses over cross-sections of a curved bar subjected to plane bending. They were first to point out the basic difference between the stress distribution in a straight and curved beam as early as 1858. Further developments of this theory were made by Grashof and Pearson. The exact solution for the particular case of stresses in curved bars of rectangular cross-section based on the integration of differential equations of the theory of elasticity was obtained by Golovin in 1881. This work was published in Russian and remained unknown in other countries until the same problem was tackled a few years later by Ribière and Prandtl (Ref. 1).

Early Experiments and Calculations

Much more thorough experimental and theoretical work dates back to the beginning of the century. In 1906 and during the following year Goodenough, Moore and others conducted experimental programs at the University of Illinois for the purpose of designing chain links (Ref. 2). However, because of the doubt regarding the distribution of pressure between adjacent links it was considered advisable to use circular rings of rectangular cross-sections loaded in diametral compression. With such elements a true knife-edge bearing was possible and the general Winkler theory could be tested without introducing unknown factors resulting from the pressure distribution. The tests on the rings, therefore, could be considered as more reliable than the link tests in establishing the truth or falsity of the analysis. The rings were loaded along a diameter and the deformations were measured by micrometers, reading directly to 0.001 and by interpolation to 0.0001 inches. Three steel rings 1 inch thick were used having outer and inner diameters of 12 and 9 inches respectively. In computing the deflections the bending and direct stresses were considered. The experiments seemed to confirm the theoretical analysis and it was concluded that the fundamental equations employed would give very closely the true stresses in rings.

An analysis of thin rings was made by Talbot in 1908 in connection with tests on cast iron and reinforced concrete pipes. He assumed that the material of the ring was homogeneous and had a constant modulus of elasticity. The theoretical investigation considered bending stresses only and deflections due to bending. It was emphasized that in case of greater thickness of the ring in comparison with the diameter, the assumptions of the ordinary theory of flexure do not hold. The length of the inner fiber is less than that of the outer fiber, a condition which modifies the analysis considerably. This type of analysis was regarded as complicated and it was pointed out that in ordinary construction the errors involved in using the simple theory of flexure would be relatively small. No experimental proof for this thesis was given.

Several years later Morley published a discussion of the engineering approach to the problem of curved beams (Ref. 3). It appeared, that until then the English and American practice was to estimate stresses in hooks by the rules applicable to straight beams and to ignore curvature. At that time Winkler's approximate theory was criticized. In Morley's opinion, however, there was not sufficient experimental evidence to indicate that the results obtained by the use of the Winkler formula were seriously in error.

Thick Ring Analysis by Timoshenko

The distribution of stresses in a thick circular ring compressed by two forces acting along a diameter was investigated by Timoshenko with the aid of the theory of elasticity (Ref. 4). In the calculations the thickness of the ring was taken equal to unity and the ratio of outer to inner diameter was 2. It was shown how tangential stresses varied at the points of the cylindrical surface along the inner diameter. The maximum normal stresses at the vertical as well as horizontal cross-sections of a ring were found to be in good agreement when calculated according to the hypothesis of plane cross-sections and the hypothesis of plane distribution of normal stresses.

Ring Tests

At about the same time Pippard and Miller published a paper concerning the stresses in links and their alteration in length (Ref. 5). For calculation of deflection the fixing couples and the thrust were expressed in terms of external force and the strain energy was taken for the entire elastic structure. Calculations and tests were made for a circular ring of 5.5 and 4.5 inches outer and inner diameter respectively. The width of the ring was 0.5 inches and the modulus of elasticity was found to be 28.5×10^6 lbs/sq. in. The agreement between the tests and calculations was described as satisfactory. In the derivation of formulas the strain energy due to bending, direct and shear stresses was considered. The neutral axis was said to be at the centroidal axis and the shear distribution factor was taken as unity.

In the early twenties more interest was generated in the design theories of curved structural members. It was considered for instance that when a circular ring was subjected to a tensional load produced by two equal and opposite forces applied to the internal boundary, the engineering approximations were satisfactory provided the ring was narrow, but ceased to be valid when the radius of the inner circle became reduced. It was held also that at the point of application of the load a concentrated transverse pressure was created. The physical meaning of such a pressure was then explained as follows. Under a concentrated load the material was intensely compressed and tended to expand laterally with great force. The effect of the expansion was considered to be purely local as long as the ring was relatively thin.

Review of Curved Beam Theory

Two years later Winslow and Edmonds reviewed state-of-the-art in curved beam theory and conducted several experiments concerning the stress distribution which was still subject to diverse opinions (Ref. 6). The tests were made to determine strain curves. The stress in radial direction was shown to be a vital factor in the design of curved beams of certain proportions. The authors reevaluated the following three theories which were then associated with curved beam design:

(1) The ordinary beam theory which assumed that in the cross-section of a curved beam the stresses due to bending were distributed according to the same law as that in the case of a straight beam.

(2) Winkler's theory which took into account the effect of curvature and was based on the hypothesis that plane transverse sections remained plane after loading.

(3) The Andrews-Pearson theory which proposed a refinement of the original Winkler theory, taking account of the additional consideration that the radial dimensions of the cross-section after loading were changed by the Poisson-ratio effect of the transverse strains resulting from normal stresses.

Winslow and Edmonds maintained that while radial strains have received certain comment by previous investigators the theoretical radial stress analysis of general application to curved beams apparently was not found in the engineering literature. They pointed out further, that although the Andrews-Pearson theory took account of

Poisson's lateral distortion due to circumferential stress, it ignored entirely any considerations of strains due to radial stress. It was suggested that these latter strains in certain shapes of cross-section could be very large and in a considerable portion of any cross-section might be in a direction opposite to the Poisson's lateral strains. The assumption of interdependent relation of radial and circumferential stresses made the problem sufficiently complex so that a direct mathematical solution did not appear to be possible. The authors suggested their own approach to the approximate computation of radial stresses in curved beams and proposed to use the Winkler formula for the circumferential stresses. Their paper aroused great interest and caused lengthy technical discussions involving leading authorities on elasticity. Finally, Winkler's theory was generally accepted despite its limitations.

Winkler-Bach Formula for Stress

It seems that the acceptance of the Winkler formula (later to be known as Winkler-Bach formula) for the analysis of circumferential stresses at any point in a curved beam, marks the end of the early phase in the development of classical design theory of curved flexural members. Despite certain limitations, the practical importance of Winkler-Bach formula remains unchallenged and its simple form is as follows:

$$S = \frac{M}{AR}\left[1 + \frac{c}{m(R+c)}\right] * \tag{1}$$

* Although meanings of symbols are given in the text immediately following this formula, meanings of symbols and dimensional units involved for this and other equations in this chapter are given at the end of the chapter.

Where S denotes the circumferential stress at a point c distance from the central axis of a transverse section of the beam at which the bending moment is M. The distance from the centroidal axis to the center of curvature of the unstressed beam is R. Finally, A denotes the area of the cross section and m is a property of the cross section given by the following equation:

$$m = -\frac{1}{A}\int \frac{c}{R+c}\,dA \tag{2}$$

Furthermore it can be shown mathematically that when R is made infinitely large, Eq. (1) gives

$$S_b = Mc/I = M/Z \tag{3}$$

Here, S_b is the bending stress and I and Z define the moment of inertia and the section modulus, respectively. Equation (3) is known as the straight-beam formula universally accepted by practical engineers and designers. It applies also with very little error to many beams that are usually regarded as curved, unless a given flexural member has relatively sharp curvature.

Castigliano Formula for Deflection

Where the amount of elastic deflection rather than stress governs the design of curved structural members probably the most convenient method for determining its value is by use of Castigliano's theorem. This extremely important theorem was first disclosed by Castigliano in 1875 and translated in 1919 by Andrews (Ref. 7). In its most general form it states that the partial derivative of the total elastic strain energy, stored in a structure, with respect to one of the forces gives the displacement of the point of application of the force in the direction of the force. The remarkable feature of this theorem is that it holds for statically determinate and redundant structures alike regardless of geometry and size of the structure as long as the method of superposition applies and the material is elastic. The terms force and displacement in the statement of the theorem have generalized meaning. For instance force may denote external load, bending couple or a twisting moment. Similarly, the term displacement may represent deflection, slope or angle of twist as the case may be. Mathematically the theorem of Castigliano may be stated as follows:

$$Y = \frac{\partial U}{\partial P} \tag{4}$$

In equation (4) U defines the total elastic strain energy stored in a load carrying member, P denotes an external force acting at any point of this structural member and Y stands for the displacement at the point of application and in the direction of the external force.

Equation (4) represents a powerful mathematical tool in structural design and analysis which, incidentally, is much more frequently employed in Europe than here in America. It applies to straight and curved structural members alike and is very useful in the problem solution of the following three categories:

    a. *The displacement is required at the point of application of a particular force and in the direction of that force.* The method of solution follows directly from the interpretation of equation (4), and is usually referred to as the application of the first principle of Castigliano.

b. *The displacement is required at a point other than that at which a given external force is applied.* Here use is made of a fictitious force, of an infinitely small value, acting at the point and in the direction of the displacement sought. In this case the total strain energy U is expressed in terms of all the real and fictitious quantities and the displacement is calculated according to equation (4). This approach is often referred to as method of unit loads. In the application of the principle of Castigliano the total energy involving real and fictitious quantities is differentiated with respect to the fictitious force. The fictitious force is then made equal to zero and the remaining expression yields the desired displacement.

c. *The magnitudes of statically indeterminate reactions are required.* In this instance the total strain energy of a structure is expressed as a function of the unknown redundant reactions and the partial derivatives of the strain energy with respect to each of the redundant reactions are set equal to zero, to obtain as many simultaneous equations as there are statically indeterminate quantities. This procedure is often known as the application of the theorem of least work or the second principle of Castigliano. The mechanics of deriving the required simultaneous equations once again follows from equation (4), with this stipulation the $\partial U/\partial P = 0$. In physical sense, $\partial U/\partial P = 0$ implies that the required statically-indeterminate quantity P, acting at a particular point, prevents that point from being displaced.

Closing Remarks

The preceding early history of curved member design, however sketchy, indicates that the original contributions of Winkler and Castigliano have certainly stood the test of time. The basic elementary formulas (1), (3) and (4) have universal appeal and use in solving machine design problems which involve the theory of curved members. The development of engineering methods has always been accompanied by more rigorous solutions of the theory of elasticity coupled with photoelastic studies. A comparatively recent review of state-of-the-art by Leeman (Ref. 8) with particular reference to analytical solutions for stresses in a circular ring, indicates that the error involved in applying the approximate Winkler-Bach theory is not significant. Furthermore, from the designer's point of view their theory is generally safe because it tends to err on the conservative side.

Symbols for Chapter 2

A	Area of cross-section, in.^{2}
c	Distance from central axis to extreme fiber, in.
I	Moment of inertia, in.^{4}
M	Bending moment, lb-in.
m	Section property in Winkler-Bach formula
P	Concentrated load, lb
R	Radius of curvature to central axis, in.
S	Stress, psi
S_b	Bending stress, psi
U	Elastic strain energy, lb-in.
Y	Deflection, in.
Z	Section modulus, in.^{3}

CHAPTER THREE

Engineering Aspects of Design

Introduction

Since almost every engineering problem has many solutions, to design is to make a practical and workable decision. This decision will influence the size, geometry and material of the finished item and it will be the product of engineering assumptions, mathematical methods of analysis and knowledge of materials application. Because of simplifying assumptions involved the design decision is usually a compromise. To make the correct technical choice a comprehensive knowledge of the theoretical design principles is needed in addition to good practical experience and creative imagination. The development of new solutions or the improvement of the existing ones also involves the theoretical methods and supporting experimental evidence. The application of the theoretical methods leads usually to a number of answers which must be subsequently selected and interpreted to assure correct physical significance of the results. It should therefore be remembered that although the knowledge of mathematical models is essential to meaningful engineering analysis, even the best formulas contain certain limitations. With the current tendency to heavier loads, higher speeds, increased machine complexity and weight restrictions, the extent of such limitations should constantly be kept in mind.

Engineering Assumptions

The simplifying engineering assumptions may be of basic or design character. Here basic assumptions are employed in the derivation of the fundamental equations while the design assumptions are concerned with simplifying a complex machine part so that it could realistically fit a mathematical model. However it is clear that although the mathematical models are as a rule, useful engineering tools, sometimes they fail to completely agree with experiments so that empirically determined correction factors may have to be applied to the theoretical equations. Many improvements have been made to date in experimental techniques and a complete discipline has evolved known as experimental mechanics. One of the more important branches of experimental mechanics, directly applicable to design, is experimental stress analysis (Ref. 9).

One way of accounting for an element of uncertainty, present in most designs, is to adopt realistic design factors which can be related to strength, stiffness or other mechanical characteristics of the part under consideration. In designing curved machine elements and similar structural members the first calculations are usually based on strength and rigidity considerations. Depending on the manner of loading and support the strength factors can be developed for bending, torsion, shear, tension, compression, fatigue resistance, creep, impact and other basic characteristics. In particular, when dealing with construction of machine tools, stiffness criterion may be of primary importance. For instance cutting force could produce excessive deflection of the work-piece and cause inaccuracies in the finished product (Ref. 10). Stiffness is also desirable for smooth performance of closely assembled parts as it decreases excessive wear and vibration amplitudes.

Use of Design Formulas

Many design formulas for curved and straight flexural members are available and therefore it will be in order to discuss here briefly the most general rules of their use with specific regard to engineering assumptions involved (Ref. 11). Although in most cases slide-rule accuracy is satisfactory in making the computations, some formulas for curved beams, rings, arched elements and complex flat springs involving the algebraic addition of quantities which are large in comparison with the final result, must be calculated to at least four significant figures. Such formulas, as shown in various chapters throughout this text, involve numerous trigonometric terms, and although they appear unnecessarily elaborate their simplification is difficult and generally leads to substantial error. If a complex formula is expected to be in frequent use, good engineering practice would be to represent it graphically.

When deriving a new design formula according to a selected mathematical model, involving uncertain boundary conditions or load distribution, it is always recommended to effect some sort of rough check on the results obtained, utilizing bracketing assumptions and analogies with careful reference to engineering fundamentals. Such a common sense approach saves time and money.

When calculating elastic deflections in curved machine members involving a combination of straight and circular portions there is frequently a tendency on the part of some designers either to ignore the effect of curvature entirely, or to make some assumption as to the equivalent length of such a structure. The basic concepts of the developed or equivalent length are described in some detail in Chapter 8. The design formulas and charts indicate that a significant error can be introduced by over simplifying assumptions applied to machine elements consisting of straight and curved portions even in the most elementary design cases.

Design Criteria

In all situations structural design criteria must depend on specific requirements. The extent of allowable permanent strain for homogeneous metal parts is usually about 0.2 percent, with the exception of local regions of stress concentration where higher plastic strains may be admissible without jeopardizing the overall performance of a machine. Rigidity requirements for machine tools, however, are usually quite stringent.

Ultimate load on a machine member may be defined as the maximum working load multiplied by a suitable factor of safety. Such a factor makes some allowance for extreme conditions dependent on loading rates, temperature cycling and other external factors the combined effect of which may be impossible to predict.

The theoretical methods of analysis are based on the assumption of linear relationship between the stress and strain. A typical stress-strain curve for a ductile structural material, such as low-carbon steel, is shown in Fig. 3-1. In this diagram the slope of the elastic portion of the curve has been exaggerated for the purpose of clarity. The linear relationship between the stress and strain was first discovered by Robert Hooke in 1678 and his law is still extensively used in most practical applications. For the case of pure tension, shaded portion of the diagram represents the elastic strain energy. This energy can be recovered upon removal of external loading. This loading produces stress S_o* and strain e_o. The total area under the stress-strain curve represents the material's toughness. Here a significant portion of the energy is dissipated as heat during the process of permanent deformation of the material.

Apparent Versus Actual Stress

The stress-strain diagram, depicted in Fig. 3-1, not only illustrates the resilience and toughness of the material but also clarifies the engineering aspects of design theory which may be sometimes overlooked. For instance, consider a machine element subjected to strain e_2, corresponding to stress S_2. The magnitude of S_2, calculated on the the assumption of Hooke's law, $e = S/E$, may be relatively high and appear totally unacceptable. In the diagram this magnitude is shown as apparent elastic stress. If we now follow the line of constant strain to the point of intersection with the stress-strain curve it is easy to see that the actual stress S_1 is only a little higher than the maximum elastic stress S_o. The difference between the actual stress S_1 and the elastic stress S_o can be tolerated provided the corresponding value of allowable permanent set, such as for instance 0.2 percent, is not exceeded. In this case therefore the design would be acceptable despite a rather high value of the calculated apparent stress. Furthermore, as stated previously, the calculated permanent strain may be so highly localized that its effect on the total structural integrity of a machine component under consideration will prove to be relatively insignificant. A rather typical example, illustrating this case, is well known to spring designers. It concerns the calculation of the maximum circumferential stress in a conical disk spring, or as it is often referred to, Belleville washer. The theoretical methods of analysis in this particular case are based essentially on the elastic behavior and the corresponding design formulas give stress values in excess of 300,000 psi, and sometimes as high as 600,000 psi. At the same time it has been observed that these washers display only limited permanent set and generally give satisfactory service. Naturally, a more accurate stress formula could be derived on the assumption of elastic-plastic behavior of this spring. A more practical alternative however would be to develop an empirical correction for the existing formulas in order to bring down the calculated stress values to a more realistic level. Such

---

* For meaning of symbols and dimensional units involved see material at end of chapter.

a procedure, could be, in effect, compared with reducing stress level S_2 down to S_1, as shown in Fig. 3-1.

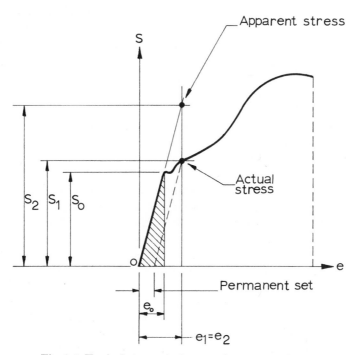

Fig. 3-1. Typical stress-strain curve for structural steel

Stress Concentration

Other pertinent illustrations of the significance of high local intensity of stress may be quoted easily from engineering experience. For instance in the design of riveted members made in structural steel the usual practice is to estimate a tensile stress, assumed to be uniformly distributed over the working cross-section. At the same time relatively high stresses, which are likely to develop at the edges of rivet holes, are ignored. This is justified again because high intensity of stress leads to a local plastic yielding which causes stress redistribution. High local intensity of stress produced by such irregularities as notches, screw threads, holes, sharp corners and similar abrupt changes in structural shape, is characteristic of elastic behavior. Hence, the practical significance of such a stress concentration will depend largely on the type of material under consideration.

By the definition, factor of stress concentration is usually expressed as the ratio of the actual maximum stress to the apparent stress calculated by the design formulas using the net cross-section of a member but ignoring the effect of form irregularities on the distribution of elastic stress.

Under static loading localized plastic stresses in ductile materials may have only limited effect on the structural integrity of a given machine part. However, the problem of such a stress concentration should not be ignored in the design of machine elements subjected to fatigue loading.

In brittle parts, stress concentration is always a major consideration. Even under static loading conditions brittle material displays the inability to mitigate stress concentration and unpredictable nature of crack propagation.

In general, stress concentration factors can be obtained mathematically or experimentally. In developing such factors by theory the assumptions of the theory of elasticity are utilized. Photoelastic and strain-gauge techniques have been particularly convenient in experimental approach. One of the more serious drawbacks in the analysis of stress concentration is the appearance of biaxial and triaxial stress systems even for the simplest conditions of loading and geometry. In this connection, various stress criteria for brittle and ductile materials have been proposed for the correlation of experimental data. In the final analysis engineering judgement should be brought into play when differentiating between the brittle and ductile characteristics of the material in question. As a rough practical guide the elongation of 5 percent may be assumed as a dividing line between the two types of structural materials.

Since the problem of stress concentration was recognized numerous investigations of theoretical and experimental nature were made during the last fifty years. A very complete information on various stress concentration factors was given by Peterson (Ref. 12). His factors are directly applicable to machine design.

Factor of Safety

The general concept of factor of safety is known to every practicing engineer. For ductile materials, such as low carbon steel, it represents usually the ratio of the yield strength of a material to the allowable working stress. The latter is often referred to as the design stress. For brittle structural materials, for which there is no defined yield point, the factor of safety is normally based on the ultimate strength. The selection of a rational value of the factor of safety is important because of its direct relation to the economy and safety of the design. For instance, when the selected factor is too high the design becomes uneconomical because the corresponding working stress is too low. On the other hand when the factor is chosen too low the relevant working stress becomes excessively high and safety of the design may be impaired. The difficulty in selecting the right value stems from the fact that the yield strength of a material must be derived from experiments while the allowable working stress is established on the basis of engineering judgement and experience. Typical influencing conditions affecting the process of selection include load variability, dynamic effects, stress concentration, material deterioration, material uniformity, working environment, method of analysis and unforeseen circumstances (Ref. 13). In general, the factor of safety may be interpreted as the product of several influencing factors (Ref. 14).

$$F = F_1 \times F_2 \times F_3 \times \ldots F_n \qquad (5)$$

The number of the influencing factors may be as high as ten or more, but for the immediate practical needs four of these factors can be introduced as being of primary inportance (Ref. 15).

$$F = F_1 \times F_2 \times F_3 \times F_4 \qquad (6)$$

The above four primary factors may be described as follows. F_1 denotes here the ratio of the ultimate strength to the elastic limit of the material. For ductile materials this ratio varies roughly between 1.5 and 2. The values of the second factor F_2 may be selected according to the following criteria:

Static load .... $F_2 = 1$

Load varying between zero and maximum .... $F_2 = 2$

Alternate tension and compression of equal
magnitude .... $F_2 = 3$

The third factor F_3 depends on the manner of load application.

Load gradually applied .... $F_3 = 1$

Load suddenly applied .... $F_3 = 2$

Impact or shock loading value for factor F_3 must be calculated for each individual case of machine part geometry, manner of loading, and material characteristics. In this case the shape of the load-time curve is of special importance. Various approximate and exact methods of stress and strain analysis for shock loading are available (Refs. 14 and 15).

The last factor, F_4, is often called the factor of ignorance which protects against unpredictable errors in manufacturing, service and materials. Its value may vary between 1.5 and 10, but seldom exceeds 3. Factors of safety for general use, taken from Machinery's Handbook as an illustration, are given in Table 3-1 (Ref. 15).

Table 3-1. Factor of Safety, F

Material	Static Load	Load varying from zero to maximum in one direction	Load varying from zero to maximum in both directions	Sudden and Impact Loads
Cast Iron	6	10	15	20
Wrought Iron	4	6	8	12
Steel	5	6	8	12
Wood	8	10	15	20
Brick	15	20	25	30

The above factors are applicable to mean rather than lower limit of material strength.

Fortunately for the designer the selection of the factor of safety can often be avoided since many national and industrial organizations develop various design specifications in which they include recommended values of the allowable working stresses.

Strength of Materials

The design of curved members for machines is concerned with the development of analytical methods and selection of optimum materials on the basis of their strength characteristics. The science of elasticity and strength of materials is treated adequately in many college and professional publications. This section touches only on those areas of materials science which are usually the subject of more frequent design inquiries.

The majority of engineering materials can be classified under the two basic categories: brittle and ductile. Extensive theoretical and experimental investigations of brittle materials led to the determination of the important effects of surface finish and size effect on the ultimate strength of these materials. Because of the presence of incipient microscopic cracks, acting as severe stress raisers on the surface of a brittle material, the amount of work necessary to initiate brittle failure is extremely small in comparison with the theoretical strain energy required to break the molecular bond. This is one of the most fundamental limitations in the development of brittle material technology.

The selection of allowable strength values for a ductile material is based on the stress-strain diagram such as that shown in Fig. 3-1. Because, as a rule, it is relatively difficult to establish a unique point on the stress-strain curve corresponding to the onset of yielding an arbitrary value of plastic strain determines the required level of yield strength. In dealing with a stress-strain curve, a portion of which is linear, the modulus of elasticity is usually defined by the slope of the linear part. The modulus of elasticity represents stiffness of a material in the elastic range and under such conditions the numerical values of the modulus in tension and compression are essentially the same. A measure of stiffness of the material in the direction normal to the elastic strain in tension or compression is the Poisson's ratio. For metallic materials Poisson's ratio may vary between 0.25 and 0.35, while for rubber materials this ratio approaches the value of 0.45 (Ref. 14).

In stressing the material throughout the elastic range the area under the elastic portion of the stress-strain curve determines the amount of resilience stored. When the stress is released the strain energy equivalent to the amount of resilience is completely recovered. Resilience helps to carry high stresses and deformations in mechanical springs and various other elements of equipment and machinery.

In designing steel members and complex structures a material with relative high ductility is generally desired because it is capable of accommodating the appreciable strains and redistributing stresses without premature fracture. The usual measure of ductility is the percentage elongation. The latest trend in engineering design is to define the ductility of a material on the basis of the ultimate load. According to Marin (Ref. 14) the magnitude of the true ductility is

$$D_e = 100 \ln (1 + e_u)$$

Here e_u denotes the nominal strain corresponding to the ultimate load as given by the stress-strain curve. Where the point of the ultimate load is difficult to determine Marin suggests alternative methods of calculation (Ref. 14).

In some areas of material selection the familiar strength to weight ratio may be of importance. Here such materials as aluminum alloys, magnesium alloys, titanium, and plastics rate highly. One of the modern developments concerns maraging steel for which the theoretical strength to weight ratio as high as 10^6 has been obtained.

In analyzing the shear properties of a given material the modulus of rigidity G can be easily determined if the corresponding modulus of elasticity in tension E and the Poisson's ratio μ, are known.

$$G = E/2(1+\mu)$$

The strength and stiffness characteristics are important in the design of machine members for torsion. For the majority of ductile materials the yield strength in shear may be assumed to be equal to 60 percent of the tensile yield. For brittle materials the ultimate shear and tensile strengths are practically equal. In designing machine members for bending yield stress values in tension or compression are recommended.

Various aspects of strength of materials summarized in this section are based on uniaxial and static stress conditions. Because of more stringent requirements of modern machinery, utilization of combined stress properties and dynamic effects in design may soon become a necessity. The detailed discussion of the developments in this area is considered to be beyond the scope of this book. For a comprehensive study of mechanical behavior of engineering materials the reader is referred to the Marin's latest work (Ref. 14).

Symbols for Chapter 3

D_e	True ductility
E	Modulus of elasticity, psi
e	General symbol for strain
e_1, e_2	Strains at stresses S_1 and S_2
e_o	Maximum elastic strain
e_u	Strain at ultimate stress
F	Resultant factor of safety
$F_1, F_2 \ldots F_n$	Component factors of safety
G	Modulus of rigidity, psi
S	Stress, psi
S_1, S_2	Stresses at strains e_1 and e_2, psi
S_o	Maximum elastic stress, psi
μ	Poisson's ratio

CHAPTER FOUR

Analysis of Stress and Deflection

Introduction

Deformation and internal stress in an elastic body, resulting from external forces, are of fundamental interest in the design of machines. The science of materials, together with the laws of statics, provides a sound basis for engineering studies at undergraduate level and serves as an aid to the recently graduated engineer making the transition from the academic environment to design office routine.

Stress Due to Flexure

Stress distribution for a straight beam with longitudinal plane of symmetry can be calculated from the elementary flexure formula, Eq. (3), provided the relevant bending moment and sectional geometry are known. For a beam of rectangular cross-section, as shown in Fig. 4-1, the moment of inertia about the central x-x axis, which in this case is the same as the neutral axis, is given by the well known equation

$$I = \frac{bh^3}{12} *$$
(7)

Since in this case distance from outer fiber to the neutral axis is $c = h/2$, it follows from Eqs. (3) and (7) that

$$S_b = 6M/bh^2 = M/Z$$
(8)

---

* For meaning of symbols and dimensional units involved for this and other equations in this chapter see material at end of chapter.

Here $Z = bh^2/6$ denotes the section modulus for a rectangular cross-section. This form of sectional property is very convenient in calculating maximum bending stresses in straight members and in curved members for which cross-sectional dimensions are small compared with the total length of the load carrying member. Values of section modulus Z are usually tabulated in engineering handbooks for various cross-sectional geometries. According to the sign convention depicted in Fig. 4-1, bending moment is positive when the top surface of the beam is in compression. Since the bending stress varies linearly its value at any depth of the beam becomes

$$S = My/I \qquad (9)$$

In the majority of design situations we deal with curved machine

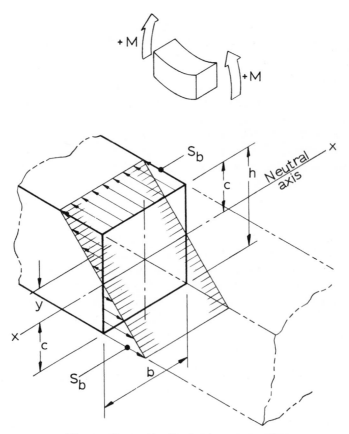

Fig. 4-1. Stress distribution in pure bending

members in which bending stress predominates. Since the design calculations start usually with the assumed or known values of the bending moment M, yield strength of the material S_y, and the relevant factor of safety F, Eq. (8) can be easily restated to give

$$Z = \frac{FM}{S_y} \tag{10}$$

For a specified width of rectangular cross-section the required depth of the beam becomes

$$h = 2.45\sqrt{FM/bS_y} \tag{11}$$

For a solid circular cross section the design formula for the diameter is

$$d = 2.17\sqrt[3]{FM/S_y} \tag{12}$$

Similar working formulas can be developed by simple algebraic transformation for other cross-section geometries.

Stress Due to Torsion

When twisting moment T is applied to the bar in a plane perpendicular to the bar axis, as shown in Fig. 4-2, the maximum torsional shearing stress can be calculated from the elementary theory of strength of materials. For a bar of diameter d this stress is

$$S_t = Td/2I_p \tag{13}$$

Since for a solid bar cross-section the polar moment of inertia is

$$I_p = \pi d^4/32$$

Substituting this polar moment into Eq. (13) gives

$$S_t = 16T/\pi d^3 \tag{14}$$

Since the modulus of rigidity, according to Hooke's law, can be expressed as the ratio of shearing stress to the shearing strain, we find that

$$G = 2LS_t/\eta d \tag{15}$$

Eliminating shearing stress between Eqs. (14) and (15) and solving for the angle of twist of the bar cross-section, shown in Fig. 4-2, yields

$$\eta = \frac{TL}{GI_p} \tag{16}$$

The general design convention for the case of twist is given in Fig. 4-2. The double arrow directed away from the given cross-section

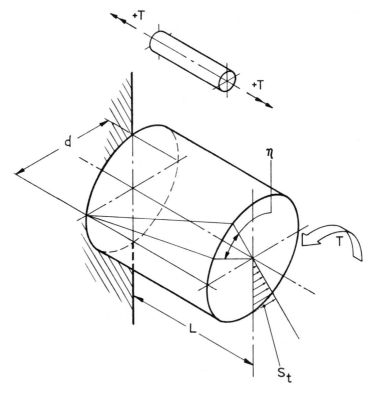

Fig. 4-2. Stress distribution in pure torsion

corresponds to the positive twisting moment T in accordance with Fig. 4-2. The shearing stress is shown to vary linearly as the distance from the center of the bar. For bars of circular cross sections, the theoretical assumption, that plane sections normal to the bar axis remain plane during the twist, holds true. However this assumption is not valid for arbitrary cross-sectional geometry.

The product term GI_p, is usally called torsional rigidity. For an arbitrary cross-section it is customary to write GK, where K may be termed torsional shape factor, depending on the geometry and dimensions of the cross section. For a circular geometry $K = I_p$. For other cross sections, K is smaller than I_p. In some cases the value of K may be only a small fraction of I_p. Further discussion of the torsional shape factor and some typical design formulas for K are given in Chapter 5.

The general formula for the maximum torsional shearing stress in an arbitrary cross-section is

$$S_t = T/K_s \qquad (17)$$

Here K_s is section modulus for torsion. For more complex geometrical shapes the equation to be used is $S_t = TC/K$, (Ref. 11) which appears to be analogous to Eq. (9) expressing the maximum flexural stress. In this case however, C is a complicated function of arbitrary cross-sectional geometry of a twisted bar. Only for a circular cross-section C may be regarded as the distance of the extreme fiber from the neutral axis. While the neutral axis represents the line of zero fiber stress in a bar subject to flexure, the torsional center is that point about which the section rotates. For many cross sectional geometries these two points do not coincide.

It should be noted that the majority of the design formulas involving torsional strength, on the assumption of straight bars, apply equally well to curved machine members provided their curvature is not unduly sharp.

Stress Due to Direct Shear

In most elementary design calculations the direct shear stress is normally defined as the shear load divided by the total cross-sectional area of a load member. Because precise knowledge of shear stress distribution is seldom available the concept of an average shear stress over the area appears to be the only practical approach to the problem. The analysis of various curved machine members, which are relatively thin, does not require the inclusion of shear stress calculations.

When the effect of shearing stress is required the distribution of this stress may be determined for simple cross-sectional shapes (Refs. 16 and 17). The transverse shear distribution for a rectangular beam cross-section, together with the relevant sign convention, is shown in Fig. 4-3. By the definition of the strength of materials theory the formula for the shearing stress in a beam derived over one hundred years ago, is

$$\tau = QJ/bI \tag{18}$$

Here Q denotes the transverse shearing force, while J is first moment of area with respect to the neutral axis, sometimes known as the statical moment. By reference to Fig. 4-3,

$$\tau = \frac{Q}{bI} \int_{y_0}^{h/2} by \, dy \tag{19}$$

Integrating Eq. (19), gives

$$\tau = \frac{Q}{2I} \left[\left(\frac{h}{2} \right)^2 - y_0^2 \right] \tag{20}$$

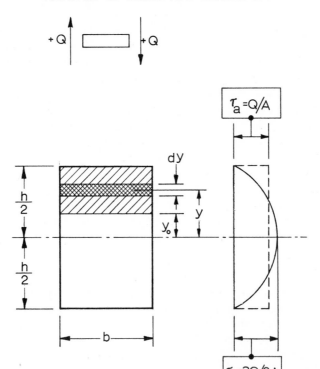

Fig. 4-3. Transverse shear distribution for a rectangular cross section

Equation (20) indicates that the shearing stress varies parabolically and that its maximum value is obtained when $y_o = 0$. It can be shown also that the shearing stress, integrated over the total area of beam cross section, is equal to the transverse shearing force Q. Finally, substituting $y_o = 0$ and Eq. (7) in Eq. (20), yields

$$\tau_m = 3Q/2A \tag{21}$$

For the case of shearing stress distribution over a circular cross-section the maximum stress becomes

$$\tau_m = 4Q/3A \tag{22}$$

For other cases, such as for instance I-beam geometry, a reasonable approximation for the maximum shearing stress may be obtained by dividing the transverse shearing force Q by the web cross sectional area. It is normally assumed that the contribution of the I-beam flanges to the transmission of the shearing force is relatively small.

Based on the above considerations, and for the purpose of further discussion, the shear distribution factor can be defined as follows: a numerical quantity which when multiplied with the average shearing stress gives the shearing stress at the centroid of the cross section. If the average shearing stress, Fig. 4-3, is denoted by τ_a and the shear distribution factor by ξ, we get

$$\tau_m = \xi \tau_a \qquad (23)$$

It may be of interest to recall here that at the upper and lower surfaces of the beam the shearing stresses vanish as indicated by Eq. (20). This is also true for other cross-sectional geometries and the above theory of strength of materials may be considered as entirely satisfactory for most practical purposes.

In the above examples the transverse shearing force Q is assumed to act through the center of twist, coinciding with the centroid of the cross section. Under such circumstances the transverse force produces no torsion of the beam. Therefore in cross-sectional areas having two axes of symmetry we have the transverse shearing stress only. For unsymmetrical cross sections the transverse shearing force causes twist of the beam around the center of twist the location of which is usually difficult to determine in relation to the centroid of the cross-section.

Torsion of Thin Walled Tubes

Many design cases involve solid cross-sections to which the elementary stress formulas given at the beginning of this chapter apply directly. In calculating shearing stresses induced by a torque in a thin-walled tube, simplified Bredt's formula can be used to advantage (Ref. 16).

$$S_t = \frac{T}{2At} \qquad (24)$$

Eq. (24) is applicable to any shape of thin walled tube cross-section. Here T denotes the twisting moment, t is the wall thickness and A defines an area enclosed by the mean contour of the wall. For instance, for a tube of circular cross-section described by mean radius r and wall thickness t, Eq. (24) becomes

$$S_t = \frac{T}{2\pi r^2 t} \qquad (25)$$

It should be noted here that A denotes the whole area described by mean radius and not the actual cross-sectional area of the tube, which in this case would be $2\pi rt$. Bredt's formula applies only to closed sections. Since the area enclosed by the circular contour is maximum for a given length of a perimeter, a circular tube is stronger and stiffer in torsion than any other form of hollow cross section, and therefore more efficient. Compare, for instance, a square thin-walled tube of mean side a and thickness t with a circular tube of the same wall thickness and weight, both subjected to the same torsional moment T, as shown in Fig. 4-4. It follows from the condition of equal weight

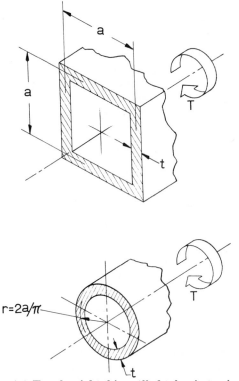

Fig. 4-4. Equal weight thin-walled tubes in torsion

that mean radius for the circular tube must be $r = 2a/\pi$. Taking the areas enclosed by the mean contours for both sections and employing Eqs. (24) and (25) we find that the shearing stress for the circular section is about 21 percent lower than that for the square tube. For the same enclosed areas, the stress is the same for both sections but the weight of the circular tube is some 11 percent lower.

Bending of Tubular Members

In calculating the flexural strength and rigidity of curved members having thin-walled tubular sections, the deformation of the cross-section under load should be considered. Von Kármán and Timoshenko contributed simplified design formulas for circular and square cross-sections, respectively (Ref. 11).

In dealing with the deflection of a circular tubular member, flexural rigidity EI should be multiplied by the following factor

$$\varphi_1 = \frac{1 + 12\varphi}{10 + 12\varphi} \tag{26}$$

where

$$\varphi = t^2 R^2 / r^2 (r + t)^2 \cong t^2 R^2 / r^3 (r + 2t) \tag{27}$$

Here R denotes mean radius of curvature, r is mean radius of tube section and t is wall thickness. It is known that comparatively thin curved tubes are more flexible in bending than the usual theory would indicate. The increased flexibility is due to flattening of the tubes during bending. Numerically, this phenomenon may be interpreted as a decrease in the moment of inertia. Hence the corresponding maximum bending stress must also be affected by the flattening of the cross-section and the relevant design formula is

$$S_b = 0.58 \, Mc / I\varphi_1 \sqrt{\varphi_2} \tag{28}$$

Here

$$\varphi_2 = \frac{6}{5 + 6\varphi} \tag{29}$$

In evaluating the deflection and stresses of a curved member, the cross-section of which is a thin hollow square, the relevant moment of inertia should be multiplied by the following factor

$$\varphi_3 = \frac{0.027 + \varphi_o}{0.066 + \varphi_o} \tag{30}$$

Where

$$\varphi_o = t^2 R^2 / a^2 (a + t)^2 \cong t^2 R^2 / a^3 (a + 2t) \tag{31}$$

Parameter φ_o, given by Eq. (31) is similar to φ given by Eq. (27). In Eq. (31), a denotes mean length of the side of the square section. Factors φ_1, φ_2 and φ_3 are given in Fig. 4-5.

In-plane Deflection of Curved Members

Elastic response of a machine member to external loading is usually judged by the magnitudes and directions of deflections. These values

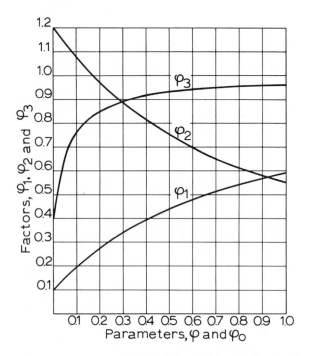

Fig. 4-5. Factors for strength and rigidity of thin tubular members

are directly related to the state of stress in a given member and are useful in correlating experimental data. The theoretical analysis of the elastic response involves various geometrical and physical parameters for which several methods of calculation are available in the literature. Despite the usual simplifying assumptions some of the analyses prove to be exceedingly complex and the engineer is required to obtain the necessary data by purely experimental means. This book presents only a few elementary methods of deflection analysis which, despite some of their limitations, have been found relatively useful in the derivation of formulas for the preliminary design.

When considering small deflections of an essentially straight bar subject to pure bending the well known differential equation of the elastic curve is

$$\frac{d^2Y}{dx^2} = \frac{M}{EI} \tag{32}$$

Equation (32) expresses the relation between the curvature, bending moment and flexural rigidity of the beam. This relation follows from the fundamental equation of bending, originally developed by Bernoulli and Euler over two hundred years ago.

In general, a relatively slender curved member subjected to in-plane external loading, undergoes radial and tangential displacements, u and v, respectively. If the normal and bending stresses at a section of the curved member are taken into account the differential equations of the relevant elastic curve (Ref. 18), are

$$\frac{u}{\varrho} + \frac{dv}{ds} = \frac{N}{AE} - \frac{M}{AE\varrho} \tag{33}$$

$$\frac{d^2u}{ds^2} + \frac{u}{\varrho^2} = -\frac{M}{EI} \tag{34}$$

Assuming that radius of curvature $\varrho = R =$ constant and considering radial deflection due to bending alone, Eq. (34) becomes

$$\frac{d^2u}{d\theta^2} + u = -MR^2/EI \tag{35}$$

Here ds is replaced by $Rd\theta$, while θ denotes the angle measured along the axis of the curved member. When the radius of curvature is constant, such as in Eq. (35), the solution can be obtained by formal integration. Similarly, Eq. (33) can be modified for constant radius of curvature.

$$\frac{dv}{d\theta} + u = (NR - M)/AE \tag{36}$$

If the bending moment M can be expressed as a function of one variable θ, radial displacement u can be obtained from Eq. (35), in terms of angle θ, external loading and flexural rigidity of the curved member. The tangential displacement v can then be found from Eq. (36) utilizing the general expression for u. Note also that for an infinitely large radius of curvature ϱ, Eq. (34), reduces to Eq. (32), for straight beams. Furthermore, using similar substitution in Eq. (33), shows that the axial extension or contraction of the beam is proportional to the normal force N and to the length of the beam, and inversely proportional to the product of the cross-sectional area and the modulus of elasticity.

$$v = \frac{Ns}{AE} \tag{37}$$

Since the uniaxial strain $e = v/s$, and the tensile stress is simply

$S = N/A$, Eq. (37), transforms into familiar equation defining Hooke's law

$$e = S/E \qquad (38)$$

Transverse Deflection of Curved Members

The equations of the elastic line for a transversely loaded curved member may be obtained by expressing curvature and twist in terms of displacement Y and the angle of twist η. The derivation of these equations is due to Bleich (Ref. 19). For a curved bar of circular curvature the relevant expressions are

$$\frac{M}{EI} = \frac{\eta}{R} - \frac{d^2Y}{R^2 d\theta^2} \qquad (39)$$

$$\frac{T}{GK} = -\left(\frac{1}{R} \frac{d\eta}{d\theta} + \frac{1}{R^2} \frac{dY}{d\theta} \right) \qquad (40)$$

It is assumed here that downward deflection is positive. The angle of twist is considered positive when measured counterclockwise.

Equations (39) and (40) can now be solved for Y and η in the following way. First differentiate both sides of Eq. (39) with respect to θ. This gives

$$\frac{1}{EI} \frac{dM}{d\theta} = \frac{1}{R} \frac{d\eta}{d\theta} - \frac{1}{R^2} \frac{d^3Y}{d\theta^3} \qquad (41)$$

Adding Eqs. (40) and (41) eliminates the angle of twist and yields

$$\frac{d^3Y}{d\theta^3} + \frac{dY}{d\theta} = -R^2 \left(\frac{1}{EI} \frac{dM}{d\theta} + \frac{T}{GK} \right) \qquad (42)$$

The angle of twist η follows from Eq. (39)

$$\eta = \frac{MR}{EI} + \frac{1}{R} \frac{d^2Y}{d\theta^2} \qquad (43)$$

The general Eqs. (42) and (43) can be solved for specific cases of transversely loaded curved members provided the relevant expressions for the bending and twisting moments as well as the boundary conditions are known. The products EI and GK represent the flexural and torsional rigidities of a particular member while R denotes the radius of curvature. Once the equation for the transverse deflection Y is established from the solution of the differential Eq. (42), the angle of twist η can be calculated from Eq. (43) by differentiating the expression for Y twice with respect to θ and substituting the result in

Eq. (43). The alternate method of solution for the angle of twist can be established by eliminating the terms involving Y from the basic Eqs. (39) and (40). This can be accomplished by first differentiating both sides of Eq. (40) with respect to θ. This gives

$$\frac{1}{GK}\frac{dT}{d\theta} = -\frac{1}{R}\frac{d^2\eta}{d\theta^2} - \frac{1}{R^2}\frac{d^2Y}{d\theta^2} \tag{44}$$

Subtracting Eq. (44) from Eq. (39), and rearranging, yields the following differential equation for the angle of twist.

$$\frac{d^2\eta}{d\theta^2} + \eta = R\left(\frac{M}{EI} - \frac{1}{GK}\frac{dT}{d\theta}\right) \tag{45}$$

The above differential equation can be solved in the usual way by utilizing the expressions for the bending and twisting moments in a particular design case with due regard to the relevant boundary conditions. It should be noted that the solution of Eqs. (42) and (45) requires a total of five boundary conditions to evaluate five integration constants. If the deflection and the angle of twist are required simultaneously it is convenient to employ Eqs. (42) and (43). An example of such a solution is presented in Chapter 11, dealing with transversely loaded, thin circular rings.

Elastic Strain Energy

The application of Castigliano principles, outlined briefly in Chapter 2, depends on the interpretation of the expressions for strain energy with regard to the conditions of loading and support. As a general guide the strain energy should be expressed in terms of statically independent forces and the integration extended over the entire load carrying member. In dealing with statically indeterminate structures the energy is expressed as a function of redundant reactions. Various aspects of this procedure are illustrated throughout the book in solving actual design problems.

According to Timoshenko (Ref. 17) the general expression for the total strain energy in a curved bar of relatively deep cross section is

$$U = \int_0^\theta \left(\frac{M^2}{2AER\delta} + \frac{N^2}{2AE} + \frac{\xi Q^2}{2AG} - \frac{MN}{AER}\right)Rd\theta \tag{46}$$

Here M, N and Q denote bending moment, normal force and shearing force, respectively. The term δ defines the distance between the neutral and central axis while ξ denotes the shear distribution factor which depends entirely on the geometry of the cross-section. Symbols

E and G denote moduli of elasticity and rigidity. Radius of curvature R, cross-sectional area A and angle θ complete the notations used in the above equation.

It is evident from Eq. (46) that the energy of strain is represented by functions of the second degree in the external forces and moments. The displacements are assumed to be proportional to the applied loads. When this is not the case the strain energy is no longer a second degree function although the material of the structure may still follow Hooke's law. Such a special case should be analyzed by other methods the description of which is considered to be beyond the scope of this treatise.

Equation (46) is applicable to in-plane loading and it can be simplified considerably for curved members whose radial thickness is small in comparison with the radius of curvature. In such cases the shift of the neutral axis becomes rather small and bending need only be considered. The relevant theory of the shift of the neutral axis δ is discussed in Chapter 12 in more detail as an introduction to the analysis of links, hooks, thick rings, curved beams and similar machine members. When thin sections only are taken into account the strain energy due to bending is

$$U = \int_0^\theta \frac{M^2 R d\theta}{2EI} \qquad (47)$$

For slender curved members loaded normal to the plane of curvature bending and twisting occur simultaneously. Essentially we are concerned here with a general three-dimensional problem. However, since loading entirely normal to the plane of curvature does not produce any deformation within this plane, the effects of the external loading acting normal to the plane of curvature only need be analyzed. The transverse bending and twisting moments in this particular case contribute to the total strain energy. This gives.

$$U = \int_0^\theta \left(\frac{M^2}{2EI} + \frac{T^2}{2GK} \right) R d\theta \qquad (48)$$

In Eq. (48), T represents the twisting moment which is not an independent quantity but a function of external loading.

Analysis of In-Plane Displacements by
Method of Castigliano

In calculating the deflection or slope at a point of an elastic member the Castigliano theorem, expressed by Eq. (4), is often convenient to apply. If the total strain energy, given by Eq. (46), can be expressed in terms of P and M_o, considered as the external force and externally applied bending couple, respectively, at which the deflection and slope are sought, the general expressions for relatively thick curved members become

$$Y = \int_0^\theta \left[\frac{M}{AER\delta}\left(\frac{\partial M}{\partial P}\right) + \frac{N}{AE}\left(\frac{\partial N}{\partial P}\right) + \frac{\xi Q}{AG}\left(\frac{\partial Q}{\partial P}\right) \right.$$
$$\left. - \frac{M}{AER}\left(\frac{\partial N}{\partial P}\right) - \frac{N}{AER}\left(\frac{\partial M}{\partial P}\right) \right] R\, d\theta \tag{49}$$

$$\psi = \int_0^\theta \left(\frac{M-N\delta}{AER\delta}\right)\left(\frac{\partial M}{\partial M_o}\right) R\, d\theta \tag{50}$$

For the case of relatively thin members, Eqs. (49) and (50) reduce to

$$Y = \int_0^\theta \frac{M}{EI}\left(\frac{\partial M}{\partial P}\right) R\, d\theta \tag{51}$$

$$\psi = \int_0^\theta \frac{M}{EI}\left(\frac{\partial M}{\partial M_o}\right) R\, d\theta \tag{52}$$

Equations (51) and (52) are simple in use and can be employed in many practical design situations. The advantage in using, for instance Eq. (51) instead of Eq. (35) lies in the fact that in solving Eq. (51), evaluation of constants of integration is entirely avoided. The results obtained with the aid of either of these equations are, of course, in complete agreement. Numerous practical examples will be solved in detail throughout various chapters of this book, utilizing the above equations, so that the mechanics of derivation of the design formulas based on Castigliano's principle will become quite familiar.

Analysis of Transverse Displacements by Method of Castigliano

Since in the majority of applications of transversely loaded curved members the effect of bending and twisting moments is of primary importance, the strain energy can be given by Eq. (48). The error due to such simplification is considered to be generally small compared with much more serious unavoidable errors due to uncertain boundary conditions encountered in such applications. It has been shown already that the deflection and the angle of twist of a transversely loaded curved member can be calculated from the differential equations (42), (43) or (45). The same results can be derived from

Eq. (48) utilizing the principle of Castigliano given in most general form by Eq. (4). Hence if the transverse displacement under load P is required, and if the bending and twisting moments M and T, are some functions of P, we get

$$\frac{\partial U}{\partial P} = \int_0^\theta \frac{M}{EI} \left(\frac{\partial M}{\partial P}\right) R \, d\theta + \int_0^\theta \frac{T}{GK} \left(\frac{\partial T}{\partial P}\right) R \, d\theta \qquad (53)$$

After calculating the deflection from Eq. (53), the slope in the plane perpendicular to the plane of curvature is found from

$$\psi = \frac{1}{R} \left(\frac{dY}{d\theta}\right) \qquad (54)$$

The angle of twist at the point of application of the externally applied twisting couple T_o can also be obtained from Eq. (48) by taking the partial derivatives with respect to T_o, considered as an independent variable. This gives

$$\frac{\partial U}{\partial T_o} = \int_0^\theta \frac{M}{EI} \left(\frac{\partial M}{\partial T_o}\right) R \, d\theta + \int_0^\theta \frac{T}{GK} \left(\frac{\partial T}{\partial T_o}\right) R \, d\theta \qquad (55)$$

In Eqs. (53) and (55), the bending and twisting moments are given as functions of P. However if the twisting couple T_o is also some function of P, the Eq. (4) becomes

$$Y = \frac{\partial U}{\partial P} + \left(\frac{\partial U}{\partial T_o}\right) \left(\frac{\partial T_o}{\partial P}\right) \qquad (56)$$

Therefore when this functional relationship between T_o and P is specified, the deflection can be found from Eq. (56) by substituting in it the relations (53) and (55) and by calculating the partial derivative of T_o with respect to P. Equation (56) finds application to some statically indeterminate problems, such as transversely loaded circular frames with built-in supports, in which T_o may represent a restraining twisting moment at one end of the frame, (Ref. 17).

Remarks on the Use of Castigliano Equations

Several design equations given in the preceding sections for deflection analysis find their origin in the general statement of Castigliano, $Y = \partial U/\partial P$. In order to clarify the mechanics of dealing with Castigliano equations, such as, for instance Eqs. (49), (50), (51) and others, consider a classical example of a uniform curved beam, forming the quadrant of a circle and carrying a concentrated vertical load P, as shown in Fig. 4-6. To calculate the upward deflection of the lower

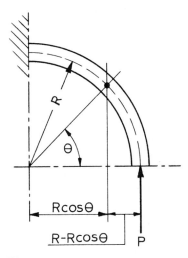

Fig. 4-6. Quarter circle curved beam

end of the beam due to bending, Eq. (51) will be used. Since the cross-section of the beam, radius of curvature and the modulus of elasticity are assumed to be constant, and since the beam subtends 90 deg angle, Eq. (51) can be written as follows:

$$Y = \frac{R}{EI} \int_0^{\pi/2} M \frac{\partial M}{\partial P} d\theta \qquad (51a)$$

It may be noted here that formal integration with respect to θ can be carried out only if the terms under the integral sign can be expressed as suitable functions of θ. To do this consider an arbitrary point defined by θ, and write the expression for the bending moment about this point, caused by the concentrated load P. The elementary trigonometric relation gives

$$M = PR \, (1 - \cos \theta) \qquad (51b)$$

In Eq. (51b), R is constant by the definition of the problem. The terms P and θ represent independent variables. According to the usual rules of calculus, $\partial M/\partial P$ implies partial differentiation of the function given by Eq. (51b). It is well to recall at this point of the discussion that the general rules for finding partial derivatives are the same as for ordinary derivatives, and that essentially the partial derivatives have the same geometric interpretation as that for the derivative of a function of one variable. Such considerations should convince the designer or student of engineering that the level of mathematics required for the solution of Castigliano-type equations is quite elementary and should encourage the reader to practice applications of this ingenious principle.

Hence the term $\partial M/\partial P$ is found easily from Eq. (51b) by holding the other independent variable θ constant and considering M to be a function of only one variable, P. This gives

$$\frac{\partial M}{\partial P} = R\,(1 - \cos\theta) \tag{51c}$$

We now have two functional relationships (51b) and (51c), expressed in terms of θ, which can be substituted directly into the Castigliano equation (51a). Therefore the next step gives

$$Y = \frac{PR^3}{EI} \int_0^{\pi/2} (1 - \cos\theta)^2\, d\theta \tag{51d}$$

The integration is performed with respect to θ, between the limits indicated according to the usual rules of calculus. First the trigonometric expression may be transformed as follows:

$$(1 - \cos\theta)^2 = \frac{3}{2} - 2\cos\theta + \frac{\cos 2\theta}{2}$$

Inserting the above expression in Eq. (51d), performing integration term by term and substituting the relevant limits gives

$$Y = \frac{PR^3}{EI} \left| \frac{3\theta}{2} - 2\sin\theta + \frac{\sin 2\theta}{4} \right|_0^{\pi/2}$$

$$Y = \left(\frac{3\pi - 8}{4}\right) \frac{PR^3}{EI}$$

or

$$Y = 0.3562\, PR^3/EI \tag{57}$$

The above detailed procedure is typical of many applications of the Castigliano equations. The trigonometric functions are very useful in dealing with such problems. The formulas for a variety of curved members presented in this book involve the algebraic addition of trigonometric terms and often require more than three significant figures in the computations. For this purpose a convenient summary of trigonometric functions is given in Table 4-1.

Table 4-1. Auxiliary Trigonometric Functions

x, deg.	15	30	45	60	75	90
x, rad	0.26180	0.52360	0.78540	1.04720	1.30900	1.57080
x^2	0.06854	0.27416	0.61685	1.09663	1.71348	2.46741
$\sin x$	0.25882	0.50000	0.70711	0.86603	0.96593	1.00000
$\cos x$	0.96593	0.86603	0.70711	0.50000	0.25882	0
$x \sin x$	0.06776	0.26180	0.55536	0.90691	1.26440	1.57080
$x \cos x$	0.25288	0.45345	0.55536	0.52360	0.33880	0
$x^2 \sin x$	0.01774	0.13708	0.43618	0.94971	1.65510	2.46741
$x^2 \cos x$	0.06620	0.23743	0.43618	0.54832	0.44348	0
$\sin 2x$	0.50000	0.86603	1.00000	0.86603	0.50000	0
$\cos 2x$	0.86603	0.50000	0	—0.50000	—0.86603	—1.00000

x, deg.	105	120	135	150	165	180
x, rad	1.83260	2.09440	2.35619	2.61799	2.87979	3.14159
x^2	3.35842	4.38651	5.55163	6.85387	8.29319	9.86959
$\sin x$	0.96593	0.86603	0.70711	0.50000	0.25882	0
$\cos x$	—0.25882	—0.50000	—0.70711	—0.86603	—0.96593	—1.00000
$x \sin x$	1.77016	1.81381	1.66609	1.30900	0.74535	0
$x \cos x$	—0.47431	—1.04720	—1.66609	—2.26726	—2.78168	—3.14159
$x^2 \sin x$	3.24400	3.79885	3.92561	3.42694	2.14644	0
$x^2 \cos x$	—0.86923	—2.19326	—3.92561	—5.93566	—8.01064	—9.84959
$\sin 2x$	—0.50000	—0.86603	—1.00000	—0.86603	—0.50000	0
$\cos 2x$	—0.86603	—0.50000	0	0.50000	0.86603	1.00000

Symbols for Chapter 4

A	Area of cross-section, in.^{2}
a	Side of square section, in.
b	Width of rectangular section, in.
C	Geometrical function for a twisted bar, in.
c	Distance from central axis to extreme fiber, in.
d	Diameter of solid bar, in.
E	Modulus of elasticity, psi

G	Modulus of rigidity, psi
h	Depth of cross-section, in.
I	Moment of inertia, in.^{4}
I_p	Polar moment of inertia, in.^{4}
J	First moment of area, in.^{3}
K	Torsional shape factor, in.^{4}
K_s	Section modulus for torsion, in.^{3}
L	Length of straight bar, in.
M	Bending moment, lb-in.
M_o	Externally applied bending couple, lb-in.
N	Normal force, lb.
P	Concentrated force, lb.
Q	Transverse shearing force, lb.
R	Radius of curvature to central axis, in.
r	Radius of circular cross-section, in.
S	Stress, psi
S_b	Bending stress, psi
S_t	Torsional stress, psi
S_y	Yield stress, psi
s	Length of curved member, in.
T	Twisting moment, lb-in.
T_o	Externally applied twisting couple, lb-in.
t	Wall thickness, in.
U	Elastic strain energy, lb-in.
u	Radial displacement, in.
v	Tangential displacement, in.
x	Arbitrary distance, in.
Y	Transverse deflection, in.

y, y_o, y_1	Distances from a reference axis, in.
Z	Section modulus, in.^{3}
δ	Distance from neutral to central axis, in.
θ	Angle at which forces are considered, rad
η	Angle of twist, rad
ϕ	Parameter in Von Kármán's equation
ϕ_1, ϕ_2	Von Kármán's factors for thin tubes
ϕ_o	Auxiliary factor
ϕ_3	Timoshenko's factor for square tubes
ψ	Slope, rad
ϱ	General symbol for radius of curvature, in.
ξ	Shear distribution factor
τ	Transverse shearing stress, psi
τ_a	Average transverse shearing stress, psi
τ_m	Maximum transverse shearing stress, psi

CHAPTER FIVE

Properties of Sections

Introduction

Fundamental properties of plane sections are always required in the calculations of strength and rigidity of machine components. The properties of interest usually include cross sectional areas, moments of inertia, sectional moduli and centroid locations. Less frequently required properties, but also of importance, are torsional shape factors, location of neutral axis relative to central axis and shear distribution factors. The relevant symbols, definitions and formulas are given in many engineering publications. The mathematics necessary for the derivation of such formulas represents a wide spectrum of difficulty and will not be repeated in this text. However, the basic concepts, some unique features, practical short-cuts and helpful data will be presented for some basic shapes of cross-sections which are more frequently employed in machine design.

Moment of Inertia

The moment of inertia, referred to often as the second moment of an area, is a mathematical statement which expresses the sum of the products found by multiplying each element of the area by the square of its distance from a given axis. This axis lies in the plane of the cross-section. The maximum and minimum moments of inertia can be usually found by inspection or by calculation depending on the complexity of the shape of the section.

The two basic principles most commonly applied in the design calculations concern the moment of inertia of composite sections and the so called parallel axis theorem.

a) The algebraic sum of the moments of inertia of component parts is equal to the moment of inertia of the complete cross-section.

b) The moment of inertia of an area with respect to a central axis attains a minimum value.

Denoting by I_o the moment of inertia with respect to an axis parallel to the central axis, the parallel axis theorem gives

$$I_o = I + Ay^2 \, *\tag{58}$$

Here A denotes the total area of the cross-section and y is the distance between the parallel and the central axes.

In order to illustrate the general procedure of developing working formulas consider for instance the moments of inertia about the central axis and about the base of a rectangular cross-section utilizing direct integration and parallel axis theorem. Take b and h as width and height of the cross-section as shown in Figs. 5-1 and 5-2.

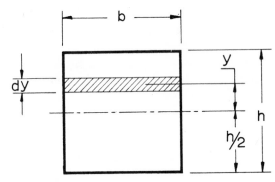

Fig. 5-1. Notation for moment of inertia about central axis of rectangle

Expressing the definition of the moment of inertia in mathematical terms, gives

$$I = \int y^2 dA\tag{59}$$

For the case of central axis, Fig. 5-1, y varies between $-h/2$ to $+h/2$, and the elementary area $dA = bdy$. Then

$$I = \int_{-h/2}^{+h/2} by^2 dy = \frac{by^3}{3}\Big|_{-h/2}^{+h/2}$$

* For meaning of symbols and dimensional units involved for this and other equations in this chapter see material at end of chapter.

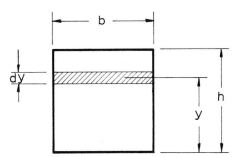

Fig. 5-2. Notation for moment of inertia about base of rectangle

which on substitution of limits, gives Eq. (7)

$$I = \frac{bh^3}{12}$$

For the case illustrated in Fig. 5-2, the general equation is the same, but the limits of integration for y are now 0 and h. This yields directly

$$I_o = \frac{bh^3}{3} \tag{60}$$

To obtain the moment of inertia about the base line of the rectangle, using parallel axis theorem, Eq. (58), write

$$I_o = \frac{bh^3}{12} + (bh) \left(\frac{h}{2}\right)^2 = \frac{bh^3}{3}$$

The moment of inertia for the cross-section of a relatively thin tubular member is often employed in the design calculations. The precise formula in this case is

$$I = \frac{\pi}{4}(R_o{}^4 - R_i{}^4) \tag{61}$$

Here R_o and R_i denote the outer and inner tube radii, respectively. Rewriting Eq. (61), gives

$$I = \frac{\pi}{4}(R_o{}^2 + R_i{}^2)(R_o + R_i)(R_o - R_i)$$

Denoting mean radius of tube by r, and wall thickness by t, as before, the above formula can be rearranged to read

$$I = \frac{\pi r t}{2}(2R_i{}^2 + 2R_i t + t^2)$$

For relatively thin tubes, term t^2 may be neglected, and making $r \cong R_i \cong R_o$, gives

$$I = \pi r^3 t \tag{62}$$

Equation (62) can also be obtained by formal integration with reference to a thin annular cross-section, using the basic expression, Eq. (59). If, for instance, the elementary area is denoted by $dA = tr d\theta$ and $y = r \sin \theta$, where θ is the angle measured from the diameter about which the moment of inertia is taken, we get

$$I = 4 \int_0^{\pi/2} tr^3 \sin^2 \theta d\theta$$

which on integration gives Eq. (62). The design formula given by Eq. (62) is much more convenient to use in the calculations than the original Eq. (61) involving fourth powers of the radii. The computation error due to the above simplification can be now analyzed. Denote the ratio of the outer to inner tube ratio by $m = R_o/R_i$. Hence the parameters of Eqs. (61) and (62) can be expressed as follows:

$$R_o{}^4 - R_i{}^4 = R_i{}^4 (m^4 - 1)$$

$$r = \frac{R_i}{2}(m + 1)$$

$$t = R_i(m - 1)$$

Dividing Eq. (61) by Eq. (62) and introducing the above relations gives

$$\frac{2(m^2 + 1)}{(m + 1)^2}$$

This parameter is illustrated graphically in Fig. 5-3.

Section Modulus

By the definition, section modulus is the moment of inertia with respect to a neutral axis divided by the distance from this axis to the most remote outer fiber. The section modulus is usually denoted by Z and largely determines the bending strength of straight and curved members of moderate curvature alike. This sectional property is extremely useful in engineering calculations and may be derived mathematically for various shapes of cross-sections. It should be noted that while there is only one value of the moment of inertia with respect to a given neutral axis, the section modulus can have two distinct values corresponding to two different distances of the extreme fibers from the neutral axis. Two different values of section modulus are therefore found in all unsymmetrical sections.

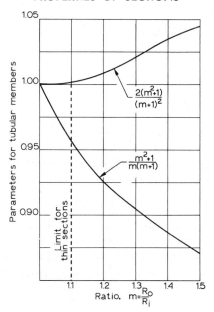

Fig. 5-3. Comparison of exact and approximate values of moment of inertia and section modulus for thin tubes

To illustrate the mathematical procedure of obtaining working formulas consider section moduli of an equilateral triangle of side length a, illustrated in Fig. 5-4.

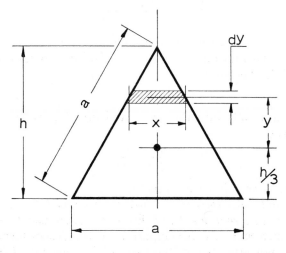

Fig. 5-4. Notation for moment of inertia about centroidal axis of triangle

Applying the general equation (59) to the centroidal axis of the triangle gives

$$I = \int_{-h/3}^{2h/3} y^2 x\, dy$$

Similar triangles yield

$$x = \frac{a}{h}\left(\frac{2h}{3} - y\right)$$

Hence substituting for x in the above moment of inertia equation, integrating and substituting the relevant limits, gives

$$I = ah^3/36 \tag{63}$$

Expressing triangle height h in terms of side length a, the minimum and maximum section-moduli follow directly from Eq. (63), by dividing the moment of inertia by the corresponding distances of the centroidal axis from the apex and base of the triangle, respectively.

$$Z_{\min} = a^3/32 \tag{64}$$

$$Z_{\max} = a^3/16 \tag{65}$$

A practical short-cut in the calculation of the section modulus for a relatively thin tubular member can be developed in a manner which will now be described.

The exact formula for the section modulus of an annular cross-section is

$$Z = \frac{\pi(R_o{}^4 - R_i{}^4)}{4R_o} \tag{66}$$

Employing the derivation procedure followed in the case of moment of inertia in the previous section, simplified design formula for the section modulus becomes

$$Z = \pi r^2 t \tag{67}$$

Dividing Eq. (66) into Eq. (67), and utilizing radius ratio m as before, we get

$$\frac{m^2 + 1}{m(m + 1)}$$

This parameter is illustrated in Fig. 5-3.

Design handbooks recommend using formulas (62) and (67) for so called very thin sections. Very thin sections are usually considered those for which $m = 1.1$. A comparison of the design formulas given in Fig. 5-3, indicates that in the case of the moment of inertia surprisingly accurate results can be obtained for a wide range of m. Even the error introduced by simplifying section modulus may be acceptable over a relatively wide range of m, considering that the accuracy with which we normally know the material properties, support and loading conditions is seldom known with certainty.

Torsional Shape Factor

In dealing with torsional rigidity of non-circular sections the standard design formulas no longer hold and it becomes necessary to introduce a factor depending on the shape and dimensions of the cross-section. In this book K denotes the torsional shape factor. There seems to be no established and generally accepted definition or symbol for this function. Such names as torsional rigidity factor, torsional resistance, torsion constant and similar have been used to denote this sectional property the dimensions of which are the same as those of moment of inertia. Some of the formulas for K, have been derived with the aid of rigorous mathematical analysis, others have been deduced through membrane analogy and are considered as approximations with not more than about 10 percent error, (Ref. 11).

The derivation of K values is generally complex because the distribution of shear stresses on a cross-section of arbitrary geometry is non-linear and the section does not remain plane under twist. Since many formulas for K are based on membrane analogy it can be concluded that narrow flanges and thin protrusions in a section have comparatively limited effect on the total torsional rigidity of a member. Also narrow sections bent into channels or open curved configurations display substantially the same torsional stiffness as that of a thin flat plate having the same length and thickness.

The value of K for a solid rectangular cross-section is of special interest because the torsional shape factor of a cross-sectional area consisting of rectangular components is roughly equal to the sum of K values of individual rectangular areas. A very useful torsion parameter K/bh^3, for a rectangular cross-section is plotted in Fig. 5-5.

Section Modulus for Torsion

In calculating torsional stresses for a member with an arbitrary cross-section, Eq. (17) can be used. In this equation, K_s, denotes a sectional property which may be called section modulus for torsion by analogy to section modulus in pure bending. Both properties Z and K_s are expressed in in.^{3} and their values are given in Table 5-1

Table 5-1. Typical Sectional Properties

Section	Section Area A	Moment of Inertia About x-x Axis I_x	Section Modulus About x-x Axis Z_x	Torsional Shape Factor K	Section Modulus For Torsion K_s
	$\dfrac{\pi d^2}{4}$	$\dfrac{\pi d^4}{64}$	$\dfrac{\pi d^3}{32}$	$\dfrac{\pi d^4}{32}$	$\dfrac{\pi d^3}{16}$
	$\pi\,(R_o{}^2 - R_i{}^2)$	$\dfrac{\pi}{4}\,(R_o{}^4 - R_i{}^4)$	$\dfrac{\pi\,(R_o{}^4 - R_i{}^4)}{4R_o}$	$\dfrac{\pi}{2}\,(R_o{}^4 - R_i{}^4)$	$\dfrac{\pi\,(R_o{}^4 - R_i{}^4)}{2R_o}$
	$2\pi r t$	$\pi r^3 t$	$\pi r^2 t$	$2\pi r^3 t$	$2\pi r^2 t$
	bh	$\dfrac{bh^3}{12}$	$\dfrac{bh^2}{6}$	For K/bh^3 See Fig. 5-5	For K_s/bh^2 See Fig. 5-5
	$bh - b_o h_o$	$\dfrac{bh^3 - b_o h_o{}^3}{12}$	$\dfrac{bh^3 - b_o h_o{}^3}{6h}$	$\dfrac{(h - h_o)\,(2b - h + h_o)^2\,(h + h_o)^2}{16\,(b + h_o)}$	$\dfrac{(h^2 - h_o{}^2)\,(2b - h + h_o)}{4}$

Table 5-1 (Continued). Typical Sectional Properties

	Section Area A	Moment of Inertia About x-x Axis I_x	Section Modulus About x-x Axis Z_x	Torsional Shape Factor K	Section Modulus For Torsion K_s
	bt	$\dfrac{bt^3}{12}$	$\dfrac{bt^2}{6}$	$\dfrac{bt^3}{3}$	$\dfrac{bt^2}{3}$
	πab	$\dfrac{\pi ab^3}{4}$	$\dfrac{\pi ab^2}{4}$	$\dfrac{\pi a^3 b^3}{a^2+b^2}$	$\dfrac{\pi ab^2}{2}$
	$\pi\,(ab - a_o b_o)$	$\dfrac{\pi\,(ab^3 - a_o b_o^3)}{4}$	$\dfrac{\pi\,(ab^3 - a_o b_o^3)}{4b}$	$\dfrac{\pi a^3 b^3}{a^2+b^2}\left[1 - \left(\dfrac{a_o}{a}\right)^4\right]$	$\dfrac{\pi ab^2}{2}\left[1 - \left(\dfrac{a_o}{a}\right)^4\right]$
	$\pi t\,(a + b)$	$\dfrac{\pi b^2 t\,(3a + b)}{4}$	$\dfrac{\pi bt\,(3a + b)}{4}$	$\dfrac{4\pi^2 abt\,(ab - at - bt)}{\Theta}$ $\Theta = \pi\,(a + b - t)\left[1 + 0.27\dfrac{(a - b)^2}{(a + b)^2}\right]$	$\pi t\,(2ab - bt - at)$
	$0.433a^2$	$0.018\,a^4$	max $a^3/16$ min $a^3/32$	$\dfrac{\sqrt{3}}{80}a^4$	$a^3/20$

for some typical configurations. Torsion parameter K_s/bh^2 for a rectangular cross-section is given in Fig. 5-5.

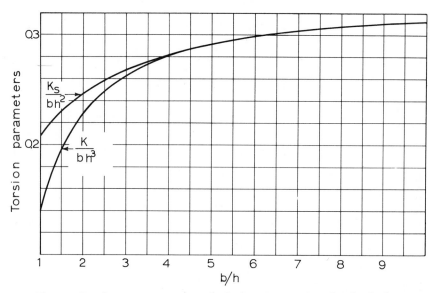

Fig. 5-5. Torsion parameters for calculating stress and angle of twist for
rectangular cross-sections

Compound Cross-Section

In many problems of machine design typical cross-sectional geometries are used for which the properties are given in design handbooks and other publications. However in calculating composite cross-sections additional procedures must be employed in order to end-up with a single design formula describing the particular property for a given cross-sectional geometry. In such situations there is usually more than one approach to the problem.

Consider for instance the moment of inertia of an elementary built-up cross-section, shown in Fig. 5-6. The usual method of finding the total moment of inertia of the composite section is to employ Eq. (58) which requires that the neutral axis of the whole section is first found. This is accomplished by breaking the whole area into component areas and finding the sum of first moments of individual component areas about a selected reference axis such as, for instance, the base line shown in Fig. 5-6. The first moment is defined as the product of the area and the distance of the centroid of this area from the

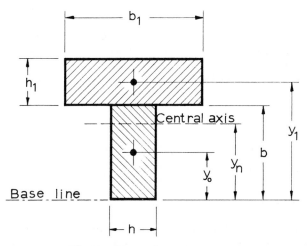

Fig. 5-6. Composite cross-section

reference axis. The sum of the first moments is then divided by the total area of the composite cross-section to give the distance of the neutral axis from the base line. With reference to Fig. 5-6, this definition gives

$$y_n = \frac{bhy_o + b_1h_1y_1}{bh + b_1h_1}$$

Also, by analogy to Eq. (59) first moment of area can be defined mathematically as follows:

$$J = \int y \, dA \qquad (68)$$

The moment of inertia of a composite cross-section can also be found more directly without calculating the neutral axis, (Ref. 20). From Eq. (58), the moment of inertia about the neutral axis is

$$I = I_o - Ay_n{}^2 \qquad (69)$$

By the definition of the first moment of area

$$y_n = J/A \qquad (70)$$

Hence substituting Eq. (70) in Eq. (69), neutral axis term is dropped out to give

$$I = I_o - \frac{J^2}{A} \qquad (71)$$

In Eq. (71), I_o stands for the sum of the two moments I_b and I_g. Here I_b denotes the product of an individual area and the distance squared measured from its center of gravity to the base line. The moment of inertia of an individual area about its own center of gravity is denoted by I_g. Hence Eq. (71) may be restated as follows:

$$I = I_b + I_g - \frac{J^2}{A} \qquad (72)$$

In some design situations I_g is relatively small and may be neglected, committing only a moderate conservative error.

Numerical Examples

Design Problem 1: Calculate the minimum moment of inertia of the composite cross-section shown in Fig. 5-6, assuming the following dimensions: $b = 2$ in., $h = 1$ in., $b_1 = 4$ in., and $h_1 = 0.5$ in.

Solution: Using symbols from Figure 5-6, Eq. (72) can be expressed as follows:

$$I = bhy_o^2 + b_1h_1y_1^2 + \frac{hb^3 + b_1h_1^3}{12} - \frac{(bhy_o + b_1h_1y_1)^2}{bh + b_1h_1}$$

Substituting the numerical data, gives

$$y_o = 1$$
$$y_1 = 2.25$$
$$bh = 2$$
$$b_1h_1 = 2$$

Hence

$$I = 2 + 2 \times 2.25^2 + \frac{1 \times 2^3 + 4 \times 0.5^3}{12} - \frac{(2 + 2 \times 2.25)^2}{2 + 2}$$

$$I = 2.27 \text{ in.}^4$$

Design Problem 2: Calculate the angle of twist and the maximum shearing stress in a cantilever member of 20 inch length, under the twisting moment $T = 2000$ lb-in. Assume the cross-sectional shape and dimensions to be the same as in Problem 1. The modulus of rigidity is $G = 12 \times 10^6$ psi.

Solution: The torsional shape factor K, for the section illustrated in Fig. 5-6 can be obtained directly from Fig. 5-5 for side ratios of the rectangular parts equal to $b/h = 2$ (vertical portion) and $b_1/h_1 = 8$ (horizontal portion). Hence

$$K/bh^3 = 0.229$$

and

$$K/b_1 h_1^3 = 0.307$$

Since the total torsional shape factor is approximately equal to the sum of the above component factors, substituting the relevant numerical values gives

$$K = 0.229 \times 2 \times 1^3 + 0.307 \times 4 \times 0.5^3$$

$$K = 0.611 \text{ in.}^4$$

Hence, utilizing Eq. (16), the angle of twist is

$$\eta = \frac{TL}{GK} = \frac{2000 \times 20}{12 \times 10^6 \times 0.611}$$

$$\eta = 0.0055 \text{ rad.}$$

Using similar procedure for the section modulus for torsion K_s, Fig. 5-5 yields

$$K_s/bh^2 = 0.246$$

$$K_s/b_1 h_1^2 = 0.307$$

Hence

$$K_s = 0.246 \times 2 \times 1^2 + 0.307 \times 4 \times 0.5^2$$

$$K_s = 0.799 \text{ in.}^3$$

The maximum torsional shearing stress is now found from Eq. (17).

$$S_t = \frac{T}{K_s} = \frac{2000}{0.799}$$

$$S_t = 2500 \text{ psi}$$

Symbols for Chapter 5

A	Area of cross-section, in.^{2}
a	Side of a triangle, in.
b, b_1	Width dimensions, in.
G	Modulus of rigidity, psi
h, h_1	Depth dimensions, in.
I	Moment of inertia, in.^{4}

I_o	Moment of inertia about an arbitrary axis, in.[4]
I_b	Moment of inertia with respect to base line, in.[4]
I_g	Moment of inertia with respect to central axis, in.[4]
J	First moment of area, in.[3]
K	Torsional shape factor, in.[4]
K_s	Section modulus for torsion, in.[3]
L	Length of straight bar, in.
$m = R_o/R_i$	Radius ratio
R_o	Outer radius of tube, in.
R_i	Inner radius of tube, in.
r	Mean radius of tube, in.
S_t	Torsional stress, psi
T	Twisting moment, lb-in.
t	Wall thickness, in.
x	Arbitrary distance, in.
y, y_o, y_1	Distances from a reference axis, in.
y_n	Distance to central axis, in.
Z	Section modulus, in.[3]
$Z_{\max}$	Maximum section modulus, in.[3]
$Z_{\min}$	Minimum section modulus, in.[3]
η	Angle of twist, rad
θ	Arbitrary angle, rad

CHAPTER SIX

In-Plane Loading of Arcuate Members

Arcuate Member Under Vertical Load

The arcuate member of uniform, thin cross-section shown in Fig. 6-1 is often met in practice. Because this structure behaves as a statically determinate beam the bending moment at the section defined by θ follows from considerations of statics. By analogy to the previously established relations depicted in Eqs. (51a), (51b) and (51c) the integration can be performed for an arbitrary angle ϕ subtended by the curved member, (Ref. 21). The actual bending moment related to Fig. 6-1 and previous sign convention established in Fig. 4-1 gives

$$M = PR \ (\cos \theta - 1)^* \tag{73}$$

Introducing ϕ as angle subtended by the curved member the bending moment at the built-in end becomes

$$M = PR \ (\cos \phi - 1) \tag{74}$$

In the majority of practical applications ϕ seldom exceeds 180 degrees. It is evident from Eq. (74) that for $\phi = \pi$ the bending moment is $M = -2PR$ and this value is not exceeded when ϕ increases beyond 180 deg. For instance when $\phi = 270$ deg, $\cos \phi = 0$ and the bending moment at the built-in end reduces to $-PR$. In the extreme case when ϕ approaches 2π, the bending moment tends to zero while the

---

* For meaning of symbols and dimensional units involved for this and other equations in this chapter see material at end of chapter.

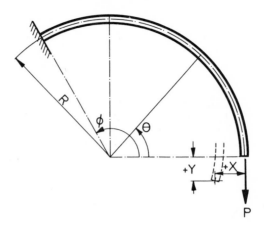

Fig. 6-1. Arcuate beam under vertical load

transverse shear at this point attains the full value of load P. The average shearing stress obtained by dividing the force P by the area of the cross-section parallel to this force is usually relatively small. Consequently the design for stress should be made only in regard to maximum bending.

In designing the member shown in Fig. 6-1 for deflection the Castigliano theorem is especially useful. According to this theorem the displacement of the free end of the arcuate beam measured in the direction of the applied load P is given by the following expression in polar coordinates:

$$Y = \frac{1}{EI} \int_0^\theta M \frac{\partial M}{\partial P} R\, d\theta \qquad (75)$$

As before EI denotes flexural rigidity. For the purpose of this derivation EI remains constant and the bending moment is given by Eq. (73). In evaluating the derivative term $\partial M/\partial P$ we observe as before that R and θ are regarded as constant. Hence employing elementary rules of differentiation, Eq. (73) gives

$$\frac{\partial M}{\partial P} = R\,(\cos \theta - 1) \qquad (76)$$

Substituting expressions (73) and (76) into the Castigliano equation (75), extending the integration over the entire curved member, and substituting the relevant limits, Eq. (75) yields

$$Y = \frac{PR^3}{4EI}(6\phi + \sin 2\phi - 8 \sin \phi) \qquad (77)$$

The vertical deflection Y is considered here positive when measured downward.

Fictitious Force Method

To find the deflection of the member shown in Fig. 6-1 at a point other than at which a given external load P is applied, use is made of the fictitious force concept together with the classical theory of Castigliano. This fictitious or imaginary force, considered to be of an infinitely small value, is applied at a point and in the direction of the required displacement. The bending moment equation is expressed in terms of all the real and imaginary quantities and total strain energy is differentiated with respect to the fictitious force selected. The fictitious force is then made equal to zero and the deflection formula obtained. To illustrate this extremely powerful design tool consider the horizontal displacement of the free end of the member shown in Fig. 6-1. If the fictitious horizontal force $\bar{H}$ is applied at the free end of this beam as shown in Fig. 6-2 then the modified bending moment equation, involving real and virtual forces, becomes

$$M = PR \ (\cos \theta - 1) - \bar{H}R \sin \theta \qquad (78)$$

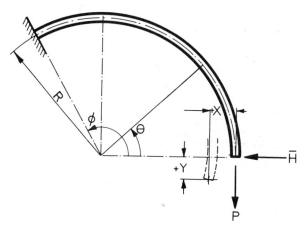

Fig. 6-2. Arcuate beam under real and fictitious loading

The partial derivative with respect to the fictitious force $\bar{H}$ follows from the usual rule of differentiating the bending moment M, given by Eq. (78) with respect to $\bar{H}$, assuming that other quantities involved such as P, R and θ remain constant. Hence, we have

$$\partial M / \partial \bar{H} = -R \sin \theta \qquad (79)$$

The required horizontal displacement can be now calculated from the following Castigliano equation:

$$X = \frac{1}{EI} \int_0^\phi M \, \frac{\partial M}{\partial \overline{H}} \, R \, d\theta \tag{80}$$

Hence substituting Eqs. (78) and (79) into Eq. (80) yields

$$X = \frac{R^3}{EI} \int_0^\phi [\overline{H} \sin^2 \theta - P \sin \theta \, (\cos \theta - 1)] \, d\theta \tag{81}$$

Since by definition the fictitious force $\overline{H}$ is considered to be extremely small compared with force P, all the terms in Eq. (81) containing $\overline{H}$ may be dropped out. Integrating the remaining terms in Eq. (81) and evaluating the limits, gives the design formula for the horizontal displacement of the arched cantilever at the point of application of the vertical load P.

$$X = \frac{PR^3}{4EI} (\cos 2\phi - 4 \cos \phi + 3) \tag{82}$$

The horizontal displacement here is assumed to be positive when measured inward towards the center of curvature of the curved member.

Arcuate Member Under Horizontal Load

The design equations for the case of a horizontal end load H, illustrated in Fig. 6-3, can be obtained by the use of Castigliano's principle and the results of the previous analysis. In applying the

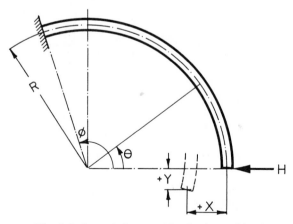

Fig. 6-3. Arcuate beam under horizontal load

results of this analysis we call on Maxwell's theorem of reciprocal deflections. In the science of mechanics of structures and materials Maxwell's theorem states that the work done on an elastic structure by a force P, in acting through the displacement produced by a force H, is equal to the work done by the force H in acting through the displacement caused by the force P. In stating this theorem the nomenclature of the Figs. 6-1 and 6-3 has been purposely retained to demonstrate its direct relation to our problem. The vertical displacement of the free end of an arcuate beam resulting from horizontal loading and depicted in Fig. 6-3, is numerically equal to the horizontal displacement due to a vertical load illustrated in Fig. 6-2. Hence we have

$$Y = \frac{HR^3}{4EI}(\cos 2\phi - 4 \cos \phi + 3) \tag{83}$$

In order to derive the design formula for horizontal displacement under a horizontal load H, as shown in Fig. 6-3, we write first the relevant bending moment equation

$$M = -HR \sin \theta \tag{84}$$

The negative sign in the above equation indicates merely that the relevant bending moment produces tensile stress on the outer surface of the arcuate beam. Differentiating the bending moment given by Eq. (84) with respect to H and substituting the result together with the bending moment expression, Eq. (84) into the general formula such as that given by Eq. (80), gives

$$X = \frac{HR^3}{EI} \int_0^\phi \sin^2 \theta \, d\theta \tag{85}$$

Integrating and substituting the limits of integration the above expression gives the following design formula for the case illustrated in Fig. 6-3

$$X = \frac{HR^3}{4EI}(2\phi - \sin 2\phi) \tag{86}$$

The deflection formulas given by Eqs. (77), (82), (83) and (86) can be now rewritten in a simplified way and developed into a set of three curves illustrating the variation of the relevant deflection coefficients with angle ϕ, subtended by the arc of the curved member under study. Hence the simplified deflection formulas become

$$Y = \frac{PR^3}{EI} K_1 \tag{87}$$

$$X = \frac{PR^3}{EI} K_2 \qquad (88)$$

$$Y = \frac{HR^3}{EI} K_2 \qquad (89)$$

$$X = \frac{HR^3}{EI} K_3 \qquad (90)$$

The deflection coefficients K_1, K_2 and K_3 are illustrated graphically in Fig. 6-4. The bending moment formulas, Eqs. (73) and (84) are relatively simple and need not be converted into graphical aids.

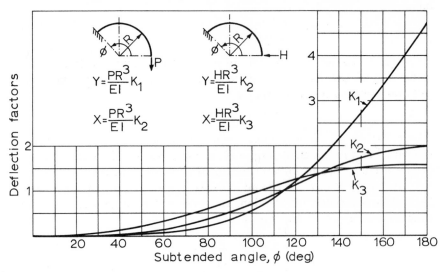

Fig. 6-4. Deflection factors for arcuate beams under end loading

Arcuate Member Under Uniform Load

Deformation of an arcuate beam carrying a circumferential uniform load shown in Fig. 6-5 can be analyzed by first considering the elementary bending moment due to unit load q. Since the elementary vertical load is $qR\, d\varepsilon$ the corresponding bending moment about a section defined by θ is

$$dM_q = qR^2 (\cos \varepsilon - \cos \theta)\, d\varepsilon$$

Integrating the above expression between 0 and θ gives

$$M_q = qR^2 (\sin \theta - \theta \cos \theta) \qquad (91)$$

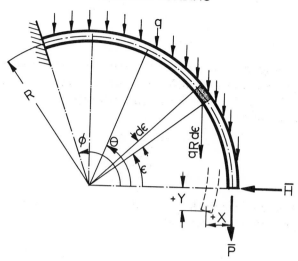

Fig. 6-5. Arcuate beam under uniform load

To find the vertical and horizontal displacements of the free end of the beam, fictitious forces $\bar{P}$ and $\bar{H}$ are applied in the directions of the required deflections as indicated in Fig. 6-5. According to the adopted sign convention, total bending moment at a section defined by θ is

$$M = qR^2 \,(\theta \cos \theta - \sin \theta) + \bar{P}R \,(\cos \theta - 1) - \bar{H}R \sin \theta \quad (92)$$

The partial derivatives necessary for the solution of the problem are identical with those given previously by Eqs. (76) and (79) because the geometry of the arcuate beam and the principal directions of the displacements at the free end remain unaltered. Hence employing the general relations Eqs. (75) and (80), based on the first theorem of Castigliano, the expressions for the derivation of the deflection formulas are

$$Y = \frac{qR^4}{EI} \int_0^\phi (\theta \cos \theta - \sin \theta) \,(\cos \theta - 1) \, d\theta \quad (93)$$

$$X = \frac{qR^4}{EI} \int_0^\phi (\sin \theta - \theta \cos \theta) \sin \theta \, d\theta \quad (94)$$

Integrating the above expressions and substituting the limits of integration, gives

$$Y = \frac{qR^4}{8EI} (2\phi \sin 2\phi - 8\phi \sin \phi + 3 \cos 2\phi - 16 \cos \phi + 2\phi^2 + 13)$$

$$(95)$$

$$X = \frac{qR^4}{8EI} (4\phi + 2\phi \cos 2\phi - 3 \sin 2\phi) \qquad (96)$$

The design formulas, Eqs. (91), (95) and (96) determine completely the maximum stresses and deflections for an arcuate member shown in Fig. 6-5. To simplify the design process, Eqs. (95) and (96) may be restated as follows:

$$Y = \frac{qR^4}{EI} K_4 \qquad (97)$$

$$X = \frac{qR^4}{EI} K_5 \qquad (98)$$

The deflection coefficients K_4 and K_5 are given in Fig. 6-6 for values of ϕ ranging between 0 and 180 deg.

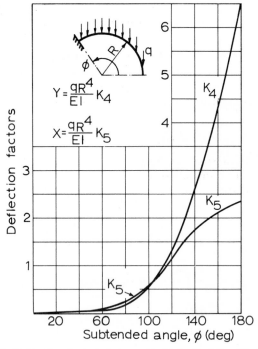

Fig. 6-6. Deflection factors for arcuate beam under uniform load

Arcuate Member Under In-Plane Couple

In some special cases of arcuate members the end load may be applied eccentrically, so that in addition to a concentrated load the effect of a bending couple must be considered. An arcuate beam subjected to an in-plane end moment is illustrated in Fig. 6-7. Here the bending moment is equal to $-M_o$ and is constant for all values of θ.

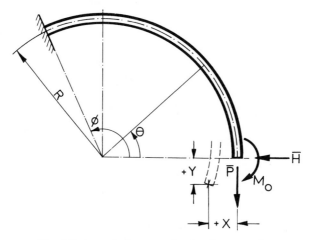

Fig. 6-7. Arcuate beam under in-plane couple

The deflection, however, varies with θ and may be determined in the following way. Consider the system of two fictitious forces $\overline{H}$ and $\overline{P}$ applied at the free end of the beam as shown in Fig. 6-7. As before these forces are considered to be infinitely small and are introduced for the purpose of establishing the direction and the location of the required deflection. Hence the bending moment equation involving real and imaginary loading on the arcuate beam is

$$M = -M_o + \overline{P}R\,(\cos \theta - 1) - \overline{H}R \sin \theta \qquad (99)$$

Since the partial derivatives with respect to the fictitious forces $\overline{P}$ and $\overline{H}$ are again the same as those given by Eqs. (76) and (79), the design formulas for deflection are obtained with the aid of the Eqs. (75) and (80).

$$Y = \frac{M_o R^2}{EI}\, K_6 \qquad (100)$$

$$X = \frac{M_o R^2}{EI}\, K_7 \qquad (101)$$

Here $K_6 = \phi - \sin \phi$ and $K_7 = 1 - \cos \phi$. The design formulas, Eqs. (100) and (101) are illustrated in Fig. 6-8.

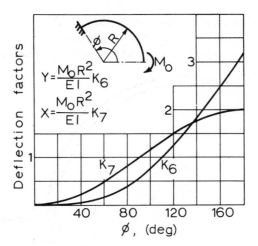

Fig. 6-8. Deflection factors for arcuate beams under in-plane end couple

General Design Formulas for Arcuate Members

For a more general case of arcuate members loaded in plane of curvature the design equations for the deflection may be derived with the aid of the theorem of Castigliano and the concept of fictitious loads as before. The general case of an arcuate member under a concentrated vertical load P, at a point at angle β from the free end is illustrated in Fig. 6-9. The bending moment equation applicable to this case is

$$M = PR \ (\cos \theta - \cos \beta) \tag{102}$$

Denoting by, α, the angle at which deflection is required, the general deflection equations become

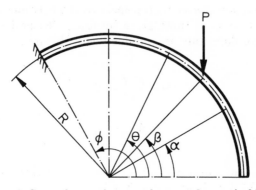

Fig. 6-9. General case of arcuate beam under vertical load

When $\alpha < \beta$

$$Y = \frac{PR^3}{EI} \left[(\phi - \beta) \cos \alpha \cos \beta - (\cos \alpha + \cos \beta)(\sin \phi - \sin \beta) \right.$$
$$\left. + 0.5 (\phi - \beta) + 0.25 (\sin 2\phi - \sin 2\beta) \right] \qquad (103)$$

$$X = \frac{PR^3}{EI} \left[\sin \alpha (\sin \phi - \sin \beta) - (\phi - \beta) \sin \alpha \cos \beta \right.$$
$$\left. + \cos \beta (\cos \beta - \cos \phi) + 0.25 (\cos 2\phi - \cos 2\beta) \right] \quad (104)$$

When $\alpha > \beta$

$$Y = \frac{PR^3}{EI} \left[(\phi - \alpha) \cos \alpha \cos \beta - (\cos \alpha + \cos \beta)(\sin \phi - \sin \alpha) \right.$$
$$\left. + 0.25 (2\phi + \sin 2\phi - 2\alpha - \sin 2\alpha) \right] \qquad (105)$$

$$X = \frac{PR^3}{EI} \left[\sin \alpha \sin \phi + \cos \beta (\cos \alpha - \cos \phi) \right.$$
$$- (\phi - \alpha) \sin \alpha \cos \beta - \sin^2 \alpha$$
$$\left. + 0.25 (\cos 2\phi - \cos 2\alpha) \right] \qquad (106)$$

It may be noted here that for $\alpha = \beta = 0$, the above five equations reduce to Eqs. (73), (77) and (82).

Employing the same nomenclature as in the above equations and referring to Fig. 6-10, the general equations for the case of a concentrated horizontal load H are found to be

$$M = HR (\sin \beta - \sin \theta) \qquad (107)$$

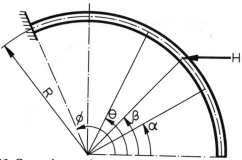

Fig. 6-10. General case of arcuate beam under horizontal load

When $\alpha < \beta$

$$Y = \frac{HR^3}{EI}\Big[0.25\,(\cos 2\phi - \cos 2\beta) + \cos \alpha\,(\cos \beta - \cos \phi)$$
$$- (\phi - \beta)\sin \beta \cos \alpha + \sin \phi \sin \beta - \sin^2 \beta\Big] \qquad (108)$$

$$X = \frac{HR^3}{EI}\Big[0.5\,(\phi - \beta) + 0.25\,(\sin 2\beta - \sin 2\phi)$$
$$- (\cos \beta - \cos \phi)\,(\sin \alpha + \sin \beta) + (\phi - \beta)\sin \alpha \sin \beta\Big]$$
$$(109)$$

When $\alpha > \beta$

$$Y = \frac{HR^3}{EI}\Big[0.25\,(\cos 2\phi - \cos 2\alpha) - (\phi - \alpha)\sin \beta \cos \alpha$$
$$+ \sin \beta\,(\sin \phi - \sin \alpha) + \cos^2 \alpha - \cos \phi \cos \alpha\Big] \qquad (110)$$

$$X = \frac{HR^3}{EI}\Big[0.5\,(\phi - \alpha) + 0.25\,(\sin 2\alpha - \sin 2\phi)$$
$$+ (\phi - \alpha)\sin \alpha \sin \beta - (\cos \alpha - \cos \phi)\,(\sin \alpha + \sin \beta)\Big]$$
$$(111)$$

For $\alpha = \beta = 0$, the above formulas simplify to Eqs. (84), (86) and (89), representing more elementary cases of arcuate beams.

For a case of in-plane loading applied eccentrically at any point of the arcuate beam, the solutions to deflection under in-plane bending moment illustrated in Fig. 6-11 may be of interest. Here the deflection formulas become

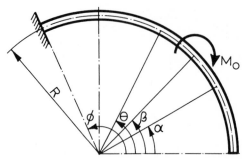

Fig. 6-11. General case of arcuate beam under in-plane couple

When $\alpha < \beta$

$$Y = \frac{M_o R^2}{EI} \left[(\phi - \beta) \cos \alpha - \sin \phi + \sin \beta \right] \qquad (112)$$

$$X = \frac{M_o R^2}{EI} \left[\cos \beta - \cos \phi - (\phi - \beta) \sin \alpha \right] \qquad (113)$$

When $\alpha > \beta$

$$Y = \frac{M_o R^2}{EI} \left[(\phi - \alpha) \cos \alpha - \sin \phi + \sin \alpha \right] \qquad (114)$$

$$X = \frac{M_o R^2}{EI} \left[\cos \alpha - \cos \phi - (\phi - \alpha) \sin \alpha \right] \qquad (115)$$

Again when $\alpha = \beta = 0$, the above formulas reduce to more elementary design Eqs. (100) and (101). The bending moment for the case shown in Fig. 6-11 is constant as before and applies only when $\theta > \beta$.

In order to determine vertical and horizontal displacements at any arbitrary point of the arcuate beam shown in Fig. 6-5, the following design equations may be used:

$$Y = \frac{qR^4}{EI} \Big[0.375 \, (\cos 2\phi - \cos 2\alpha) + 2 \cos \alpha \, (\cos \alpha - \cos \phi)$$

$$+ \, 0.25 \, (\phi^2 - \alpha^2) - \phi \sin \phi \cos \alpha + 0.25 \, (\phi \sin 2\phi + \alpha \sin 2\alpha) \Big]$$
$$(116)$$

$$X = \frac{qR^4}{EI} \Big[0.5 \, (\phi - \alpha) + 0.25 \, (\phi \cos 2\phi - \alpha \cos 2\alpha) - 0.375 \sin 2\phi$$

$$- \, 0.625 \sin 2\alpha + \sin \alpha \, (2 \cos \phi + \phi \sin \phi - \alpha \sin \alpha) \Big] \qquad (117)$$

When $\alpha = 0$, is substituted in Eqs. (116) and (117) the design formulas given by Eqs. (95) and (96) are readily obtained.

Numerical Examples

Design Problem 3: A cantilever spring of rectangular cross-section is formed in a circular arc and subtends 90 deg as shown in Fig. 6-12. External load of 5 lb acts along the line inclined at 30 deg to the horizontal axis. It is desired to calculate the maximum bending stress and vertical displacement of the free end of the spring, assuming the following data:

Mean radius of curvature	$R = 3$ in.
Width of cross-section	$b = 1$ in.
Depth of cross-section	$h = 0.05$ in.
Modulus of elasticity	$E = 30 \times 10^6$ psi

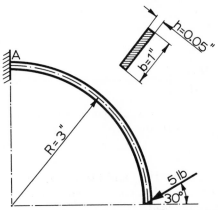

Fig. 6-12. Quarter-circle cantilever spring

Solution: Introducing $\theta = 90$ deg. into Eqs. (73) and (84) and utilizing the principle of superposition the maximum bending moment is

$$M = -R\ (P + H)$$
$$M = -3\ (5 \sin 30° + 5 \cos 30°)$$
$$M = -20.5\ \text{lb-in.}$$

Section modulus of rectangular cross-section is

$$Z = bh^2/6$$
$$Z = 1 \times 0.05^2/6$$
$$Z = 4.17 \times 10^{-4}\ \text{in.}^3$$

Hence the maximum bending stress becomes

$$S_b = M/Z$$
$$S_b = -\,20.5 \times 10^4/4.17$$
$$S_b = -49{,}200\ \text{psi}$$

The negative sign indicates that the outer surface at point A, shown in Fig. 6-12 is in tension.

The resultant vertical deflection due to the vertical and horizontal components in agreement with Fig. 6-12 is obtained from Eqs. (87) and (89) by superposition. This gives

$$Y = \frac{R^3}{EI}(PK_1 + HK_2)$$

In our case the values of K_1 and K_2 can be obtained either from Eqs. (77) and (83) or by direct reference to the design curves depicted in Fig. 6-4. Hence for $\phi = 90$ deg the corresponding deflection factors become

$$K_1 = 0.355$$
$$K_2 = 0.500$$

The moment of inertia of a rectangular cross-section of the cantilever spring is

$$I = bh^3/12$$
$$I = 1 \times 0.05^3/12$$
$$I = 10.4 \times 10^{-6} \text{ in.}^4$$

Hence substituting the numerical data into the above formula for resultant deflection Y yields

$$Y = \frac{3^3 \times 10^6 \, (5 \sin 30° \times 0.355 + 5 \cos 30° \times 0.5)}{30 \times 10^6 \times 10.4}$$
$$Y = 0.264 \text{ in.}$$

Design Problem 4: An arcuate beam of uniform rectangular cross-section is formed in a semi-circular arc as shown in Fig. 6-13. Deter-

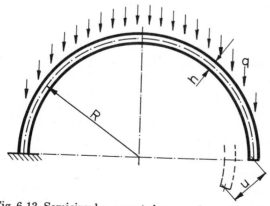

Fig. 6-13. Semicircular arcuate beam under uniform load

mine the displacement of the free end of the beam under its own weight and show that the displacement is independent of the width of the cross-section. Mean radius of curvature is 5 inches and thickness of beam amounts to 0.1 in. Take modulus of elasticity to be 30×10^6 psi and the specific weight of the material equal to 0.283 lb/in³.

Solution: If b shall denote the width of beam cross-section, the term common to both Eqs. (97) and (98) is

$$\frac{qR^4}{EI} = \frac{\pi Rbh\gamma}{\pi R} \times \frac{R^4}{E} \times \frac{12}{bh^3} = \frac{12\gamma R^4}{Eh^2}$$

Hence the required deflection is independent of the width of the cross-section.

The coefficients K_4 and K_5 can be obtained either from Eqs. (95) and (96) or the design curves given in Fig. 6-6. Hence for $\phi = 180$ deg, the deflection coefficients are

$$K_4 = 6.46$$
$$K_5 = 2.36$$

The resultant displacement, u, follows from simple addition of two vectors represented by Eqs. (97) and (98)

$$u = \frac{12\gamma R^4}{Eh^2}\sqrt{K_4{}^2 + K_5{}^2}$$

Substituting the numerical values gives

$$u = \frac{12 \times 0.283 \times 5^4}{30 \times 10^6 \times 0.01}\sqrt{6.46^2 + 2.36^2}$$

$$u = 0.049 \text{ in.}$$

In many practical applications the deflection due to the beam's own weight may be ignored. However the above formula indicates that this deflection is directly proportional to the fourth power of the radius of curvature and inversely proportional to the square of material thickness. Hence any change in these dimensions can increase or decrease the deflection rather rapidly.

Design Problem 5: Employing the method of superposition derive the design formulas for a quarter-circle arcuate beam subjected to end load P, applied eccentrically, as shown in Fig 6-14. Consider the horizontal extension bracket to be very rigid compared with the arcuate beam and express the vertical and horizontal deflections of

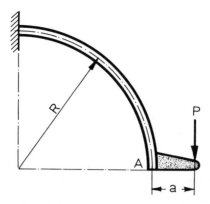

Fig. 6-14. Quarter-circle arcuate beam under eccentric end load

the free end of the beam in general symbols. Write the formula for the maximum bending stress in terms of specified horizontal deflection assuming rectangular beam cross-section.

Solution: The equilibrium of the structure, shown in Fig. 6-14, will be unaltered if by any elementary rules of statics the end load P is transferred to point A, and provided an in-plane moment $M_o = Pa$ is added at this point in order to simulate the effect of original load eccentricity. The deflection of the arcuate beam at point A becomes therefore a sum of the deflections due to end load P and in-plane end moment Pa. Hence for a case of vertical displacement, combining formulas, Eqs. (77) and (100) gives

$$Y = \frac{PR^2}{4EI} \left[2\phi \left(3R + 2a \right) - 4 \sin \phi \left(2R + a \right) + R \sin 2\phi \right]$$

When $\phi = \pi/2$, the above formula reduces to a simple expression

$$Y = \frac{PR^2}{EI} \left(0.36R + 0.57a \right)$$

Similarly, combining Eqs. (82) and (101) and substituting $\phi = \pi/2$, the formula for the horizontal displacement of the free end of the arcuate beam, subjected to eccentric loading, becomes

$$X = \frac{PR^2}{EI} \left(0.5R + a \right)$$

It may be of interest to note that the ratio of vertical to horizontal deflection in this design case is

$$Y/X = \frac{0.36R + 0.57a}{0.5R + a}$$

The above expression indicates that the horizontal displacement under vertical loading at $\phi = \pi/2$ is always greater than the corresponding vertical displacement. The maximum value of Y/X occurs when $a = 0$, which is the case of zero eccentricity. The corresponding minimum value of Y/X is about 0.57; however this value has limited practical significance. The maximum bending stress according to the elementary theory of beam flexure is

$$S_b = \frac{6P\,(R+a)}{bh^2}$$

Eliminating load term P between the formula for horizontal displacement and the above stress formula gives

$$S_b = \frac{XEh\,(R+a)}{R^2\,(R+2a)}$$

For $a = 0$, which is the case of zero eccentricity, the above stress formula becomes

$$S_b = \frac{XEh}{R^2}$$

Symbols for Chapter 6

a	Offset, in.
b	Width of rectangular section, in.
E	Modulus of elasticity, psi
H	Horizontal load, lb
$\overline{H}$	Fictitious horizontal load, lb
h	Depth of cross-section, in.
I	Moment of inertia, in.[4]
$K_1, K_2 \ldots K_n$	Deflection coefficients for inplane loading on arcuate beams
M	Bending moment, lb-in.
M_o	Externally applied bending couple, lb- in.
M_q	Bending moment due to uniform load, lb-in.
P	Vertical load, lb
$\overline{P}$	Fictitious vertical load, lb

q	Uniform load, lb/in.
R	Mean radius of curvature, in.
S_b	Bending stress, psi
u	Resultant deflection, in.
X	Horizontal deflection, in.
Y	Vertical deflection, in.
Z	Section modulus, in.^{3}
α	Angle at which deflection is required, rad
β	Angle at which load is applied, rad
γ	Specific weight of material, lb/in.^{3}
ε	Auxiliary angle, rad
θ	Angle at which forces are considered, rad
ϕ	Angle subtended by curved member, rad

CHAPTER SEVEN

Transversely Loaded
Arcuate Members

Assumptions and Sign Convention

This chapter furnishes load-deflection relations for the following cases: end load, end twisting moment, and uniformly distributed load acting normal to the plane of curvature of the bar. The method of the derivation of the design formulas for the transverse displacement, slope and angle of twist of bar cross-section is demonstrated in some detail, and the design curves are suitable for the preliminary engineering calculations, (Ref. 22).

As in the case of other curved members discussed in Chapter 6 the derived equations are valid within the elastic range of the material and apply to relatively thin curved elements of uniform cross-section.

The relevant sign convention for a curved element loaded normal to the plane of curvature is given in Fig. 7-1. All the bending and twisting moments acting normal to this plane are denoted by vectors with double arrows. The bending moments are considered positive when they produce compression on the upper surface of the beam. The twisting moment is regarded positive when it causes counter-clockwise twist of the section as shown in the sketch.

Arcuate Member Under Out-of-Plane Load

Consider the displacement characteristics of the arcuate beam

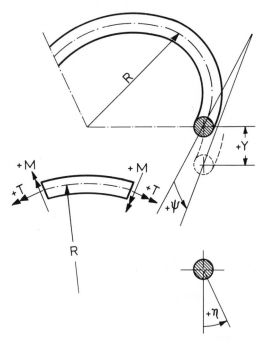

Fig. 7-1. Sign convention for transversely loaded arcuate beams

shown in Fig. 7-2. In order to calculate the displacements at a location defined by angle α, measured from the free end of the beam, a system of two fictitious forces $\bar{P}$ and $\bar{T}_o$ is assumed to exist at that

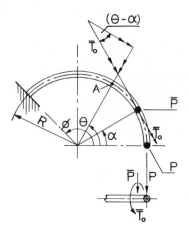

Fig. 7-2. Arcuate beam under out-of-plane concentrated load

point. The resulting bending and twisting moments at the point A can be obtained from the considerations of statics and the geometry of the curved member. It is evident from Fig. 7-2, that, theoretically, each of the forces, whether real or imaginary, contribute to the bending and twisting moments. The effect of the fictitious twisting moment $\overline{T}_o$, applied at angle α, can be obtained by resolving the vector, representing the twisting moment $\overline{T}_o$, into the normal and tangential components at A. It may be noted that this vector is tangential at the point of application of $\overline{T}_o$ and is denoted, together with its components, by double arrows. Hence following the sign convention illustrated in Fig. 7-1, the bending and twisting moments become

$$M = -P R \sin \theta - \overline{P} R \sin (\theta - \alpha) + \overline{T}_o \sin (\theta - \alpha)* \quad (118)$$

$$T = P R (1 - \cos \theta) + \overline{P} R [1 - \cos (\theta - \alpha)] + \overline{T}_o \cos (\theta - \alpha) \quad (119)$$

In equations (118) and (119) $\overline{P}$ and $\overline{T}_o$ are fictitious quantities introduced for mathematical reasons. Applying the principle of Castigliano to the transverse displacement and the angle of twist for the curved bar shown in Fig. 7-2, yields

$$Y = \frac{R}{EI} \int_\alpha^\phi M \frac{\partial M}{\partial \overline{P}} \, d\theta + \frac{R}{GK} \int_\alpha^\phi T \frac{\partial T}{\partial \overline{P}} d\theta \quad (120)$$

$$\eta = \frac{R}{EI} \int_\alpha^\phi M \frac{\partial M}{\partial \overline{T}_o} \, d\theta + \frac{R}{GK} \int_\alpha^\phi T \frac{\partial T}{\partial \overline{T}_o} \, d\theta \quad (121)$$

Equations (120) and (121) follow directly from the previously noted Eqs. (53) and (55). The bending slope at any point of the curved member can be calculated with the aid of Eq. (120), and the relation similar to Eq. (54).

$$\psi = - dY/R \, d\alpha \quad (122)$$

The partial differentiation of Eqs. (118) and (119) with respect to the ficticious quantities gives

$$\frac{\partial M}{\partial \overline{P}} = -R \sin (\theta - \alpha) \quad (123)$$

$$\frac{\partial T}{\partial \overline{P}} = R [1 - \cos (\alpha - \theta)] \quad (124)$$

$$\frac{\partial M}{\partial \overline{T}_o} = \sin (\theta - \alpha) \quad (125)$$

* For meaning of symbols and dimensional units involved for this and other equations in this chapter see material at end of chapter.

$$\frac{\partial T}{\partial \overline{T}_o} = \cos (\theta - \alpha) \tag{126}$$

It may be of interest to note that the limits of integration in Eqs. (120) and (121) are α and ϕ and not 0 and ϕ as might have been expected. The mathematical reason for this is that the partial derivatives in the interval from 0 to α must vanish since the fictitious loading $\overline{P}$ and $\overline{T}_o$ does not contribute to the bending and twisting moments when $\theta \leqq \alpha$. This is quite evident from the diagram in Fig. 7-2.

Introducing Eqs. (118), (119), (123) and (124) into Eq. (120), integrating and substituting the relevant limits gives the general expression for the displacement. Hence making the fictitious load equal to zero yields the formula for the deflection due to the end load P.

$$Y = \frac{PR^3}{EI} [0.5 \, (\phi - \alpha) \cos \alpha + 0.25 \sin \alpha - 0.25 \sin (2\phi - \alpha)]$$

$$+ \frac{PR^3}{GK} [0.75 \sin \alpha + 0.25 \sin (2\phi - \alpha) + 0.5 \, (\phi - \alpha) \cos \alpha$$

$$+ \phi - \alpha - \sin \phi - \sin (\phi - \alpha)] \tag{127}$$

Here ϕ denotes the angle subtended by the curved member and α defines the angle at which the displacement is sought.

The general expression for the slope is found next by differentiating Eq. (127) with respect to α, and substituting the result in Eq. (122). This procedure gives

$$\psi = \frac{PR^2}{EI} [0.25 \cos (2\phi - \alpha) - 0.5 \, (\phi - \alpha) \sin \alpha - 0.25 \cos \alpha]$$

$$+ \frac{PR^2}{GK} [\cos (\phi - \alpha) - 0.25 \cos (2\phi - \alpha) - 0.5 \, (\phi - \alpha) \sin \alpha$$

$$+ 0.25 \cos \alpha - 1] \tag{128}$$

To obtain the angle of twist of the curved bar shown in Fig. 7-2, in accordance with sign convention of Fig. 7-1, the formulas, Eqs. (118) and (119) together with the partial derivatives, Eqs. (125) and (126) are substituted in Eq. (121). Integrating Eq. (121) and neglecting all the terms involving $\overline{T}_o$, yields

$$\eta = \frac{PR^2}{EI} [0.25 \sin (2\phi - \alpha) - 0.5 \, (\phi - \alpha) \cos \alpha - 0.25 \sin \alpha]$$

$$+ \frac{PR^2}{GK} [\cos (\phi - \alpha) - 0.25 \cos (2\phi - \alpha) - 0.5 \, (\phi - \alpha) \sin \alpha$$

$$+ 0.25 \cos \alpha - 1] \tag{129}$$

The maximum bending and twisting moments for the calculation of stresses follow directly from Eqs. (118) and (119), and should be examined individually depending on the angle subtended by the curved member.

Arcuate Member Under Out-Of-Plane Couple

In considering the displacements of the arcuate beam carrying twisting moment at the free end, as shown in Fig. 7-3, the general calculation procedure is similar to that outlined in the previous case. Hence resolving the vectors of the real and imaginary twisting moments into normal and tangential components at A, (Fig. 7-3) as well as including the terms due to the imaginary load applied at α, the expressions for the bending and twisting components become

$$M = -\,\bar{P}R \sin{(\theta - \alpha)} + \bar{T}_o \sin{(\theta - \alpha)} + T_o \sin{\theta} \qquad (130)$$

$$T = \bar{P}R\,[1 - \cos{(\theta - \alpha)}] + \bar{T}_o \cos{(\theta - \alpha)} + T_o \cos{\theta} \qquad (131)$$

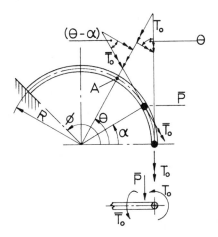

Fig. 7-3. Arcuate beam under out-of-plane couple

The partial derivatives required for the solution of this problem are identical with those given by Eqs. (123) through (126). Hence introducing Eqs. (130), (131), (123) and (124) into Eq. (120), integrating, and making the fictitious load $\bar{P}$ equal to zero, gives

$$Y = \frac{T_oR^2}{EI}\,[0.25 \sin{(2\phi - \alpha)} - 0.5\,(\phi - \alpha) \cos{\alpha} - 0.25 \sin{\alpha}]$$

$$+ \frac{T_oR^2}{GK}\,[\sin{\phi} - 0.75 \sin{\alpha} - 0.25 \sin{(2\phi - \alpha)}$$

$$- 0.5\,(\phi - \alpha) \cos{\alpha}] \qquad (132)$$

The general formula for slope can be found by differentiating Eq. (132) with respect to α and substituting this derivative in

$\psi = \dfrac{1}{R} \dfrac{dY}{d\alpha}$. This yields

$$\psi = \frac{T_o R}{EI} \left[0.25 \cos \alpha + 0.5 \left(\phi - \alpha \right) \sin \alpha - 0.25 \cos \left(2\phi - \alpha \right) \right]$$

$$+ \frac{T_o R}{GK} \left[0.25 \cos \left(2\phi - \alpha \right) + 0.5 \left(\phi - \alpha \right) \sin \alpha - 0.25 \cos \alpha \right]$$

$$(133)$$

To calculate next the angle of twist of the bar shown in Fig. 7-3, Eqs. (130), (131), (125) and (126) are substituted in Eq. (121). Integrating Eq. (121) and cancelling all the terms involving $\overline{T}_o$, yields

$$\eta = \frac{T_o R}{EI} \left[0.25 \sin \alpha + 0.5 \left(\phi - \alpha \right) \cos \alpha - 0.25 \sin \left(2\phi - \alpha \right) \right]$$

$$+ \frac{T_o R}{GK} \left[0.5 \left(\phi - \alpha \right) \cos \alpha + 0.25 \sin \left(2\phi - \alpha \right) - 0.25 \sin \alpha \right]$$

$$(134)$$

The maximum bending and twisting moments for stress calculations can be obtained from Eqs. (130) and (131) for each particular case under consideration.

Arcuate Member Under Out-Of-Plane Uniform Load

In order to develop the general displacement formulas for the case of uniform load, illustrated in Fig. 7-4, it is first necessary to calculate the bending and twisting moments due to the uniform load q alone. Consider a section at angle $\varepsilon < \theta$ and the elementary bending moment about an arbitrary point A.

$$dM_q = - \left(qR \, d\varepsilon \right) \times R \sin \left(\theta - \varepsilon \right)$$

Hence the bending moment at θ due to the uniform load is

$$M_q = - \int_0^\theta qR^2 \sin \left(\theta - \varepsilon \right) d\varepsilon$$

Integrating the above expression gives

$$M_q = -qR^2 \left(1 - \cos \theta \right) \qquad (135)$$

Similarly, the elementary twisting moment about A is

$$dT_q = qR^2 \left[1 - \cos \left(\theta - \varepsilon \right) \right] d\varepsilon$$

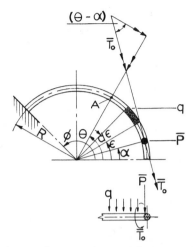

Fig. 7-4. Arcuate beam under out-of-plane uniform load

Integrating between the limits of 0 and θ yields

$$T_q = qR^2\,(\theta - \sin\theta) \tag{136}$$

Hence introducing the fictitious forces $\overline{P}$ and $\overline{T}_o$ as before and utilizing Eqs. (135) and (136) we find the total bending and twisting moments in accordance with the sign convention and notation given in Fig. 7-1.

$$M = -\overline{P}R\sin(\theta - \alpha) - qR^2\,(1 - \cos\theta) + \overline{T}_o\sin(\theta - \alpha) \tag{137}$$

$$T = \overline{P}R\,[1 - \cos(\theta - \alpha)] + qR^2\,(\theta - \sin\theta) + \overline{T}_o\cos(\theta - \alpha) \tag{138}$$

Introducing Eqs. (137), (138), (123) and (124) into Eq. (120), integrating and neglecting the terms involving fictitious loads, yields the following general expression for the transverse deflection:

$$\begin{aligned}
Y = \frac{qR^4}{EI}&\Big[1 + 0.25\cos(2\phi - \alpha) - \cos(\phi - \alpha) - 0.25\cos\alpha \\
&+ 0.5(\phi - \alpha)\sin\alpha\Big] + \frac{qR^4}{GK}\Big[1 - 0.25\cos(2\phi - \alpha) \\
&- \cos(\phi - \alpha) + \cos\phi + 0.5(\phi^2 - \alpha^2) - \phi\sin(\phi - \alpha) \\
&- 0.75\cos\alpha + 0.5(\phi - \alpha)\sin\alpha\Big]
\end{aligned} \tag{139}$$

In order to calculate the slope, Eq. (139) is differentiated with respect to α and substituted in Eq. (122). This gives

$$\psi = \frac{qR^3}{EI} [0.25 \sin \alpha + \sin (\phi - \alpha) - 0.5 (\phi - \alpha) \cos \alpha$$

$$- 0.25 \sin (2\phi - \alpha)]$$

$$+ \frac{qR^3}{GK} [\sin (\phi - \alpha) + 0.25 \sin (2\phi - \alpha)$$

$$- \phi \cos (\phi - \alpha) - 0.25 \sin \alpha - 0.5 (\phi - \alpha) \cos \alpha + \alpha]$$

$$(140)$$

The angle of twist of the arcuate beam shown in Fig. 7-4 is obtained by substituting Eqs. (137), (138), (125) and (126) into Eq. (121) and integrating between α and ϕ.

$$\eta = \frac{qR^3}{EI} [\cos (\phi - \alpha) - 0.25 \cos (2\phi - \alpha) - 0.5 (\phi - \alpha) \sin \alpha$$

$$+ 0.25 \cos \alpha - 1]$$

$$+ \frac{qR^3}{GK} [0.25 \cos (2\phi - \alpha) + \cos (\phi - \alpha)$$

$$+ \phi \sin (\phi - \alpha) - 0.25 \cos \alpha - 0.5 (\phi - \alpha) \sin \alpha - 1]$$

$$(141)$$

The calculation of the bending and torsional stresses for an arcuate beam illustrated in Fig. 7-4 can be performed with the aid of Eqs. (135) and (136).

Simplified Design of Transversely Loaded Arcuate Members

Various stress and deflection problems, involving arcuate members loaded normal to the plane of curvature, can be solved readily with the aid of the derived equations. These equations are sufficiently general and may be used in the parametric studies or preparation of graphical aids. However, since many design situations encountered in practice are limited to calculations of deflection of the free end of the beam or prediction of stresses, the general equations may be simplified by putting $\alpha = 0$. Table 7-1 presents a summary of such equations. It may be noted here that the expressions for deflection consist of two separate terms indicating the contribution of flexure and torsion, respectively. The terms involving the rigidity ratio λ, illustrate relative effect of the twist on displacement characteristics of transversely loaded arcuate beams. Separation of the flexural and torsional contributions in the design formulas is convenient because it offers the designer an early opportunity for making the selection of cross-sectional geometry with due regard to the rigidity ratio.

It is evident from the summary of formulas given in Table 7-1, that the rigidity ratio λ, influences the magnitude as well as the direction of a particular displacement. It is also noted that in certain specific cases the design formulas from Table 7-1 can be simplified further where regular, compact cross-sections are involved. For instance, for metal structures of circular cross-section $\lambda = 1.25$, and for solid square bar λ is approximately equal to 1.5. For other typical sections K and I values, required for the determination of λ, can be obtained from Chapter 5. The deflection factors B_1 through B_9 can be calculated from Table 7-2. These factors are illustrated graphically in Figs. 7-5 and 7-6.

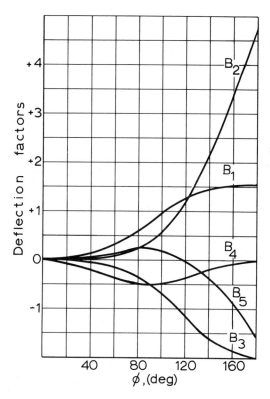

Fig. 7-5. Deflection factors for transversely loaded arcuate beams

Table 7-1. Summary of Design Equations for Transversely
Loaded Arcuate Members

Concentrated End Load

$$Y = \frac{PR^3}{EI}(B_1 + \lambda B_2)$$

$$\psi = \frac{PR^2}{EI}(\lambda B_3 + B_4)$$

$$\eta = \frac{PR^2}{EI}(\lambda B_3 - B_1)$$

$$M = -PR \sin \theta$$

$$T = PR\,(1 - \cos \theta)$$

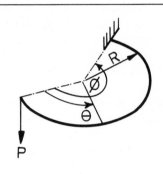

End Twisting Moment

$$Y = \frac{T_o R^2}{EI}\,(\lambda B_5 - B_1)$$

$$\psi = \frac{T_o R}{EI}\,(\lambda - 1)\,B_4$$

$$\eta = \frac{T_o R}{EI}\,(\lambda B_6 + B_1)$$

$$M = T_o \sin \theta$$

$$T = T_o \cos \theta$$

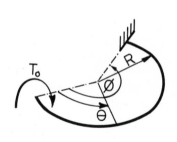

Uniform Loading

$$Y = \frac{qR^4}{EI}\,(\lambda B_7 - B_3)$$

$$\psi = \frac{qR^3}{EI}\,(\lambda B_8 + B_5)$$

$$\eta = \frac{qR^3}{EI}\,(\lambda B_9 + B_3)$$

$$M = -qR^2\,(1 - \cos \theta)$$

$$T = qR^2\,(\theta - \sin \theta)$$

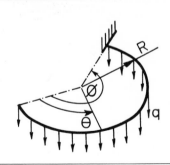

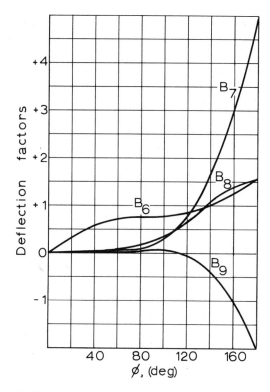

Fig. 7-6. Deflection factors for transversely loaded arcuate beams

Table 7-2. Deflection Factors for Transversely Loaded Arcuate Members

$B_1 = (2\phi - \sin 2\phi)/4$

$B_2 = (6\phi + \sin 2\phi - 8 \sin \phi)/4$

$B_3 = (4 \cos \phi - \cos 2\phi - 3)/4$

$B_4 = (\cos 2\phi - 1)/4$

$B_5 = (4 \sin \phi - \sin 2\phi - 2\phi)/4$

$B_6 = (2\phi + \sin 2\phi)/4$

$B_7 = (2\phi^2 - \cos 2\phi - 4\phi \sin \phi + 1)/4$

$B_8 = (4 \sin \phi + \sin 2\phi - 4\phi \cos \phi - 2\phi)/4$

$B_9 = (4 \cos \phi + \cos 2\phi + 4\phi \sin \phi - 5)/4$

Numerical Examples

Design Problem 6: A cylindrical bar of 1.2 inches diameter is made in the form of an arcuate beam of mean radius of curvature $R = 20$

inches. The curved bar subtends 135 deg and carries a twisting couple $T_o = 1000$ lb-in. at the free end, acting in a diametral plane normal to the plane of curvature as shown in Fig. 7-7. Calculate the vertical displacement and the angle of twist of the free end of the beam and find the stresses at the fixed end, assuming $E = 30 \times 10^6$ psi and $G = 0.4 \, E$. (Adapted from Ref. 22).

Solution: The required moment of inertia is

$$I = \frac{\pi \times d^4}{64}$$

$$I = \frac{3.1416 \times 1.2^4}{64}$$

$$I = 0.1018 \text{ in.}^4$$

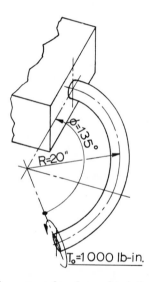

Fig. 7-7. Arcuate rod under end twisting moment

Since the torsional shape factor K for a circular cross-section is actually twice the value for I, we get

$$\lambda = \frac{EI}{GK} = \frac{EI}{0.4E \times 2I} = 1.25$$

For a value of $\phi = 135$ deg, the deflection factors B_1 and B_5 are computed with the aid of Table 7-2.

$$B_1 = 1.4281$$
$$B_5 = -0.2210$$

The equation for the vertical deflection is obtained from Table 7-1.

$$Y = \frac{T_o R^2}{EI} (\lambda B_5 - B_1)$$

$$Y = -1.7043\ T_o R^2 / EI$$

$$Y = -1.7043 \times 1000 \times 400/30 \times 10^6 \times 0.1018$$

$$Y = -0.223 \text{ in.}$$

The negative sign indicates in accordance with Fig. 7-1 that the beam is deflected upwards.

The required formula for the angle of twist is found in Table 7-1.

$$\eta = \frac{T_o R}{EI} (\lambda B_6 + B_1)$$

Again for $\phi = 135$ deg, $B_1 = 1.4281$ and $B_6 = 0.9281$. Hence the above formula reduces to

$$\eta = 2.5882\ T_o R / EI$$

By comparing directly the absolute values of the deflection and the angle of twist obtained from formulas used in this example, we get

$$\eta = \frac{2.5882}{1.7043} \times \frac{Y}{R}$$

$$\eta = \frac{2.5882 \times 0.223}{1.7043 \times 20}$$

$$\eta = 0.0170 \text{ rad}$$

The angle of twist is found to be positive and indicates counterclockwise rotation in agreement with sign convention illustrated in Fig. 7-1. The bending and twisting moments at the fixed end follow from Table 7-1.

$$M = T_o \sin \theta$$

$$M = 1000 \times 0.7071$$

$$M = 707.1 \text{ lb-in.}$$

$$T = T_o \cos \theta$$

$$T = 1000 \times (-0.7071)$$

$$T = -707.1 \text{ lb-in.}$$

Hence the bending and torsional stresses become

$$S_b = \frac{Md}{2I}$$

$$S_b = \frac{707.1 \times 1.2}{2 \times 0.1018}$$

$$S_b = 4{,}170 \text{ psi (with top surface compression)}$$

$$S_t = \frac{Td}{4I}$$

$$S_t = \frac{-707.1 \times 1.2}{4 \times 0.1018}$$

$$S_t = -2085 \text{ psi (shear in clockwise direction)}$$

Design Problem 7: The arcuate beam of hollow circular cross section is built-in at one end and subtends the angle of 180 deg, as shown in Fig. 7-8. The free end supports a vertical load $P = 50$ lb which acts

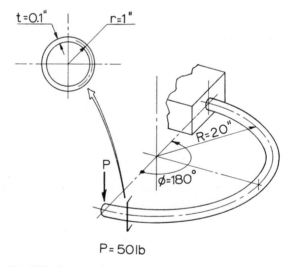

P = 50 lb

Fig. 7-8. Tubular arcuate member under transverse end load

normal to the plane of curvature. Mean radii of bend and tube-cross-section are $R = 20$ in. and $r = 1$ in., respectively. The material is steel for which $E = 30 \times 10^6$ psi and $G = 0.4\,E$. Assuming that the tube thickness is $t = 0.1$ in., calculate the transverse deflection and the angle of twist of the free end of the beam allowing for the effect of tube flattening on displacement.

Solution: According to Von Kármán's theory, referred to in Chapter 4, Eqs. (26) and (27), the factor taking into account tube flattening can be calculated from the following expression:

$$\varphi_1 = 1 - \frac{9}{10 + 12t^2R^2/r^4}$$

$$\varphi_1 = 1 - \frac{9}{10 + 12\,(0.01 \times 400)}$$

$$\varphi_1 = 0.845$$

For tubular members as well as solid circular sections and $G = 0.4\,E$, the ratio of flexural to torsional rigidity, $\lambda = 1.25$. The formulas for the required displacements are obtained from Table 7-1. Hence introducing Von Kármán's factor, gives

$$Y = \frac{PR^3}{EI\varphi_1}\,(B_1 + \lambda B_2)$$

$$\eta = \frac{PR^2}{EI\varphi_1}\,(\lambda B_3 - B_1)$$

or by substitution:

$$\eta = \frac{(\lambda B_3 - B_1)}{(B_1 + \lambda B_2)}\left(\frac{Y}{R}\right)$$

From the design Table 7-2 the deflection factors, for $\phi = 180$ deg, are

$$B_1 = 1.57$$
$$B_2 = 4.71$$
$$B_3 = -2.00$$

Taking the simplified formula for the moment of inertia of a relatively thin tube gives

$$I = \pi r^3 t$$

$$I = 3.14 \times 0.1 = 0.314 \text{ in.}^4$$

Substituting the numerical data into the above deflection formula, yields

$$Y = \frac{50 \times 20^3\,(1.57 + 1.25 \times 4.71)}{30 \times 10^6 \times 0.314 \times 0.845}$$

$$Y = 0.38 \text{ in.}$$

The required angle of twist is therefore

$$\eta = \frac{(-1.25 \times 2.00 - 1.57)}{(1.57 + 1.25 \times 4.71)} \times \frac{0.38}{20}$$

$$\eta = -0.01 \text{ rad}$$

In agreement with sign convention adopted in Fig. 7-1 negative η indicates clockwise rotation of end cross-section.

Design Problem 8: A steel rod of uniform circular cross-section is formed into a circular arc of radius R and subtends 270 deg. One end of the rod is fixed so that the rod forms an arcuate member in a horizontal plane, as shown in Fig. 7-9. If a concentrated load $P = 5$ lb is applied at the free end in the direction normal to the plane of curvature of the rod, calculate the bending and torsional stresses at sections A, B and C. Assume the diameter of rod cross-section to be 0.2 inches, and radius of curvature $R = 5$ in.

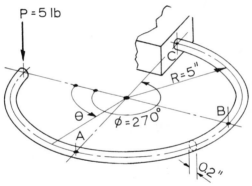

Fig. 7-9. Arcuate rod under transverse end load

Solution: The bending and twisting moments for the case considered are

$$M = -PR \sin \theta$$

$$T = PR (1 - \cos \theta)$$

According to the elementary theory of strength of materials, $S = Mc/I$, the stress equations become

$$S_b = -\frac{PR \, d \sin \theta}{2I} = -10.18 \, PR \sin \theta / d^3$$

$$S_t = \frac{PR \, d \, (1 - \cos \theta)}{4I} = 5.09 \, PR \, (1 - \cos \theta) / d^3$$

Here the moment of inertia $I = \pi \, d^4/64$ and θ denotes an angle at which the stresses are considered. On substituting the relevant numerical values of P, R and d, we get

$$S_b = -31{,}800 \sin \theta$$
$$S_t = 15{,}900 \, (1 - \cos \theta)$$

Hence the stresses are

		S_b (psi)	S_t (psi)
A	$\theta = 90°$	$-31{,}800$	$15{,}900$
B	$\theta = 180°$	0	$31{,}800$
C	$\theta = 270°$	$31{,}800$	$15{,}900$

Symbols for Chapter 7

$B_1, B_2, \ldots B_9$	Deflection factors
d	Diameter of solid bar, in.
E	Modulus of elasticity, psi
G	Modulus of rigidity, psi
I	Moment of inertia, in.^{4}
K	Torsional shape factor, in.^{4}
M	Bending moment, lb-in.
M_o	Externally applied bending couple, lb-in.
M_q	Bending moment due to uniform load, lb-in.
P	Vertical load, lb
$\overline{P}$	Fictitious vertical load, lb
q	Uniform load, lb/in.
R	Mean radius of curvature, in.
r	Mean radius of tube, in.
S_b	Bending stress, lb-in.
S_t	Torsional stress, lb-in.
T	Twisting moment, lb-in.
T_o	Externally applied twisting couple, lb-in.

$\overline{T}_o$	Fictitious twisting couple, lb-in.
T_q	Twisting moment due to uniform load, lb-in.
t	Wall thickness, in.
X	Horizontal deflection, in.
Y	Vertical deflection, in.
α	Angle at which deflection is required, rad
β	Angle at which load is applied, rad
ε	Auxiliary angle, rad
η	Angle of twist, rad
θ	Angle at which forces are considered, rad
$\lambda = \dfrac{EI}{GK}$	Ratio of flexural to torsional rigidity
ϕ	Angle subtended by curved member, rad
φ_1	Von Kármán's factor for deflection
ψ	Slope, rad

CHAPTER EIGHT

Design of Arched and
Curved-end Cantilevers

Developed Length Concept

In calculating the deflections of arched cantilevers the design people sometimes employ the concept of a developed length in conjunction with the usual formula for a straight cantilever beam, $Y = PL^3/3EI$, thereby neglecting the true effect of the curvature. It will be shown here by analysis that this concept can lead to serious errors for certain proportions of arched cantilevers and similar components.

With reference to a general case of an arched cantilever subjected to end forces and moments, shown in Fig. 8-1, the bending moment

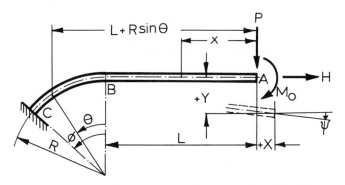

Fig. 8-1. General case of arched cantilever under end loading

acting at any point along the straight portion AB, is

$$M_1 = -Px - M_o*$$ (142)

Negative signs in the above equation indicate simply the assumption that P and M_o produce tension on the upper surface of the beam and that their effects are directly additive. The bending moment at any point defined by θ along the curved portion becomes

$$M_2 = -PR\ (k + \sin \theta) - HR\ (1 - \cos \theta) - M_o$$ (143)

Here a convenient non-dimensional ratio, $k = L/R$, is introduced. The bending moment Eqs. (142) and (143) are most general and can be used in the development of the stress and deflection formulas employing the theorem of Castigliano. It may be recalled that in order to avoid some confusion in dealing with several load terms the symbols assigned to fictitious forces carried superscript vertical bars to separate these forces from real components. This of course is a mathematical convenience only and the superscripts may be omitted provided the general procedure of derivation is kept unchanged. For instance the basic Castigliano equation for the horizontal displacement due to the transverse load P shown in Fig. 8-1 may be stated without superscript notation as follows:

$$X = \frac{1}{EI} \int_0^L M_1 \frac{\partial M_1}{\partial H}\ dx + \frac{1}{EI} \int_0^\phi M_2 \frac{\partial M_2}{\partial H} R\ d\theta$$ (144)

After obtaining $\dfrac{\partial M_1}{\partial H}$ and $\dfrac{\partial M_2}{\partial H}$ from Eqs. (142) and (143) respectively, the terms in the bending moment equations involving H and M_o are made equal to zero and the remaining terms are integrated in accordance with the above expression. For the actual case illustrated in Fig. 8-1, only the second term of the above Castigliano equation need be included in the analysis since the force H does not produce bending in the straight portion, that is $\dfrac{\partial M_1}{\partial H} = 0$.

Hence referring to Eq. (143)

$$\frac{\partial M_2}{\partial H} = -R\ (1 - \cos \theta)$$

and the equation for the horizontal deflection becomes

$$X = \frac{PR^3}{EI} \int_0^\phi (k + \sin \theta)\ (1 - \cos \theta)\ d\theta$$ (145)

* For meaning of symbols and dimensional units involved for this and other equations in this chapter see material at end of chapter.

Integrating Eq. (145) and substituting the relevant limits, gives

$$X = \frac{PR^3}{EI} (k\phi - k \sin \phi - \cos \phi + 0.25 \cos 2\phi + 0.75) \quad (146)$$

The vertical displacement under load P is obtained by first taking the partial derivatives with respect to P from Eqs. (142) and (143). This gives

$$\frac{\partial M_1}{\partial P} = -x \quad (147)$$

$$\frac{\partial M_2}{\partial P} = -R (k + \sin \theta) \quad (148)$$

Hence utilizing Eqs. (147) and (148) the relevant Castigliano equation becomes

$$Y = \frac{P}{EI} \int_0^L x^2 \, dx + \frac{PR^3}{EI} \int_0^\phi (k + \sin \theta)^2 \, d\theta$$

Integrating the above expression between the limits indicated yields

$$Y = \frac{PR^3}{EI} (0.33 \, k^3 + k^2\phi + 2k - 2k \cos \phi + 0.5\phi - 0.25 \sin 2\phi)$$

$$(149)$$

In obtaining the design formula for bending slope at the free end of the arched cantilever we find that the partial derivatives of the corresponding moments with respect to external couple M_o are $\dfrac{\partial M_1}{\partial M_o}$

$= \dfrac{\partial M_2}{\partial M_o} = -1$. Hence the equation for the slope is

$$\psi = \frac{P}{EI} \int_0^L x \, dx + \frac{PR^2}{EI} \int_0^\phi (k + \sin \theta) \, d\theta$$

which on integration gives

$$\psi = \frac{PR^2}{EI} (0.5k^2 + k\phi - \cos \phi + 1) \quad (150)$$

In considering the concept of the developed length, sometimes employed in error in design of arched cantilevers, Fig. 8-2 illustrates the equivalent total length or the projected straight length assumed for such calculations. Let us denote the equivalent length of a straight cantilever by L_e. In order to reduce the number of variables in this

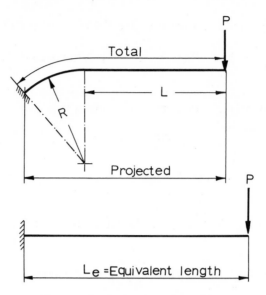

Fig. 8-2. Equivalent length cantilever

discussion, we consider the subtended angle ϕ, shown in Fig. 8-1, to be 90 deg. Hence Eq. (149) gives

$$Y = \frac{PR^3}{12EI} (4k^3 + 6\pi k^2 + 24k + 3\pi) \qquad (151)$$

The conventional deflection formula for a straight cantilever beam is

$$Y_c = \frac{PL_e^3}{3EI} \qquad (152)$$

The total equivalent length for arched cantilever with 90 deg bend is

$$L_e = L + \frac{\pi R}{2} = \frac{R}{2} (2k + \pi) \qquad (153)$$

Substituting Eq. (153) into the cantilever formula Eq. (152) gives

$$Y_c = \frac{PR^3 (2k + \pi)^3}{24\,EI} \qquad (154)$$

Dividing Eq. (154) by Eq. (151) yields

$$Y_c/Y = \frac{(2k + \pi)^3}{(8k^3 + 12\pi k^2 + 48k + 6\pi)} \qquad (155)$$

When k tends to zero, the ratio given by Eq. (155) attains the maximum value of $\pi^2/6 = 1.64$. For a bend angle of 180 deg, Eq. (149), gives

$$Y = \frac{PR^3}{6EI} \ (2k^3 + 6\pi k^2 + 24k + 3\pi) \qquad (156)$$

Similarly, the equivalent total length L_e and the deflection of a straight cantilever beam Y_c, become

$$L_e = R \ (k + \pi) \qquad (157)$$

and

$$Y_c = \frac{PR^3 \ (k + \pi)^3}{3EI} \qquad (158)$$

Hence dividing Eq. (158) by Eq. (156), gives

$$Y_c/Y = \frac{2 \ (k + \pi)^3}{2k^3 + 6\pi k^2 + 24k + 3\pi} \qquad (159)$$

When k becomes zero, ratio Y_c/Y attains the maximum value of $2 \ \pi^2/3 = 6.57$. Here the increase in subtended angle ϕ of 100 percent causes an increase of about 300 percent in deflection ratio. Therefore for relatively small values of k and for subtended angle ϕ larger than 90 deg the overestimate of the deflection for an arched cantilever may be exceedingly high. This overestimate is the result of the assumption of the developed length equal to the sum of the lengths for straight and curved portions.

The calculation of the deflection for the arched cantilevers with the aid of the approximate cantilever formula requires the use of the developed length correction factors and design curves such as those given by Palm and Thomas (Ref. 23). Their curves depict the correction factors as a function of length to radius ratio for various angles subtended by the arched cantilevers.

The assumption of the projected length in the calculations, such as shown in Fig. 8-2, is also in error and actually leads to significant underestimate. By following an identical procedure with that employed in analyzing the assumption of the total equivalent length, the deflection ratios for $\phi = 90$ deg and $\phi = 180$ deg in the case of the projected length concept are found to be

$$\phi = 90 \text{ deg} \qquad \frac{Y_c}{Y} = \frac{4 \ (1 + k)^3}{(4k^3 + 6\pi k^2 + 24k + 3\pi)} \qquad (160)$$

$$\phi = 180 \text{ deg} \qquad \frac{Y_c}{Y} = \frac{2 \ (1 + k)^3}{(2k^3 + 6\pi k^2 + 24k + 3\pi)} \qquad (161)$$

For $k = 0$, the minimum values of the ratio, obtained from formulas, Eqs. (160) and (161), are 0.42 and 0.21, respectively. Functions depicted by Eqs. (155), (159), (160) and (161) are illustrated in Figs. 8-3 and 8-4. These diagrams show the importance of including the true effect of the curvature of such machine elements when estimating the deflections and stresses.

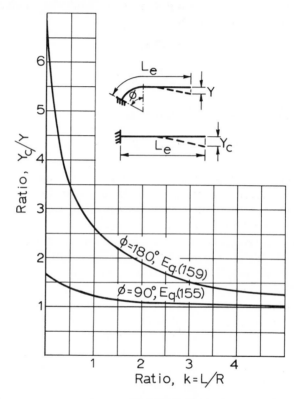

Fig. 8-3. Overestimate of calculated deflection based on stretched-out equivalent length of arched cantilever

Formulas for Maximum Deflection of Arched Cantilevers

The analysis of arched cantilever beams, subjected to end forces, may be of interest in the study of redundant frames, piping or engine lines subjected to the movement of one support of a line, for instance, with respect to the other support because of temperature gradients or assembly misalignments. If such a frame or line is restricted against any movement of supports by the adjacent parts of a machine or

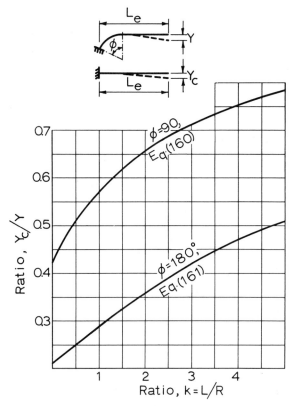

Fig. 8-4. Underestimate of calculated deflection based on projected equivalent length of arched cantilever

structure under consideration, damaging stresses can be induced in the frame or line. The knowledge of the flexibility characteristics of such parts at the stage of the preliminary design is therefore of special significance. The relevant deflection formulas can be conveniently, derived with the aid of the Castigliano method (Ref. 24).

In general, if one end of the arched cantilever such as that defined by A in Fig. 8-1 is displaced to some other position A', in order to bring this point back to A, a system of forces P, H and M_o should be applied compatible with the displacements Y, X and ψ. Equations (146), (149) and (150) express the deflections and slope of the free end of the arched cantilever caused by the transverse force P. Using the Castigliano equations together with the concept of the fictitious forces, the corresponding displacements due to the end load H are found accordingly.

$$Y = \frac{HR^3}{EI}(k\phi - \cos\phi - k\sin\phi + 0.25\cos 2\phi + 0.75) \quad (162)$$

$$X = \frac{HR^3}{EI}(1.5\phi - 2\sin\phi + 0.25\sin 2\phi) \quad (163)$$

$$\psi = \frac{HR^2}{EI}(\phi - \sin\phi) \quad (164)$$

Finally, the displacements due to the moment M_o applied at the end of the cantilever shown in Fig. 8-1, are as follows:

$$Y = \frac{M_oR^2}{EI}(0.5k^2 + k\phi - \cos\phi + 1) \quad (165)$$

$$X = \frac{M_oR^2}{EI}(\phi - \sin\phi) \quad (166)$$

$$\psi = \frac{M_oR}{EI}(k + \phi) \quad (167)$$

The design formulas, Eqs. (146), (149), (150) and Eqs. (162) through (167) inclusive describe the displacements of the free end of the cantilever under the system of forces P, H and M_o, as shown in Fig. 8-1. These equations may be used in the study of statically determinate or indeterminate structures. To further simplify the design calculations for more common geometries of arched cantilevers, graphical aids are provided for subtended angle $\phi = 90$ deg and $\phi = 180$ deg with the length to radius ratios k varying between 0 and 5. The charts and simplified formulas for these conditions are given in Figs. 8-5 and 8-6 which may be superimposed to represent other, more complex loading situations.

In some design situations the curved portion may be found at the free rather than at the built-in end of the structure, as shown in Figs. 8-7 and 8-8. These machine elements may be referred to as curved-end cantilevers.

Curved-end Cantilever with Quarter-circle Bend

Consider first the quarter-circle cantilever shown in Fig. 8-7. The bending moment at an arbitrary point on the center line of the curved portion is

$$M_1 = -PR(1 - \cos\theta) - HR\sin\theta \quad (168)$$

This bending moment equation is valid for the curved part only because of the basic difference in the geometry between AB and BC

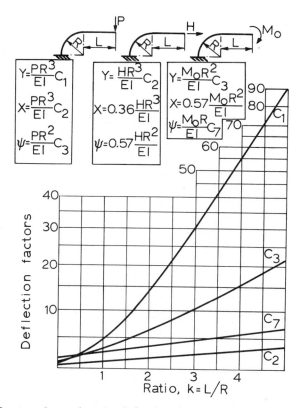

Fig. 8-5. Charts and equations for deflection of quarter-circle arched cantilevers

portions. When $\theta = \pi/2$, Eq. (168) gives the bending moment at section B.

$$M_B = -R\,(P + H)$$

The transfer of forces P and H to point B can be now effected directly so that the bending moment at a station defined by x becomes

$$M_2 = M_B - Px = -R\,(P + H) - Px \qquad (169)$$

Since we are dealing with relatively thin members the bending strain energy only is taken into account and therefore the term H should not appear as a function of x. We make the assumption here that the force H is not sufficiently large to produce such compression effects on the straight part which would markedly affect the maximum stresses and deformation of the curved-end cantilever.

With this proviso in mind we can now apply the Castigliano theorem

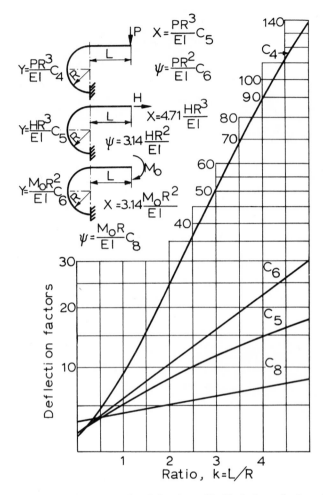

Fig. 8-6. Charts and equations for deflection of half-circle arched cantilevers

to develop the design formulas for deflection. The general expressions for this purpose are

$$Y = \frac{1}{EI} \int_0^{\pi/2} M_1 \frac{\partial M_1}{\partial P} R \, d\theta + \frac{1}{EI} \int_0^L M_2 \frac{\partial M_2}{\partial P} \, dx \qquad (170)$$

and

$$X = \frac{1}{EI} \int_0^{\pi/2} M_1 \frac{\partial M_1}{\partial H} R \, d\theta + \frac{1}{EI} \int_0^L M_2 \frac{\partial M_2}{\partial H} \, dx \qquad (171)$$

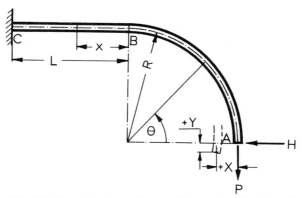

Fig. 8-7. Curved-end cantilever with quarter-circle bend

since

$$\frac{\partial M_1}{\partial P} = -R\,(1 - \cos\theta) \tag{172}$$

and

$$\frac{\partial M_2}{\partial P} = -(x + R) \tag{173}$$

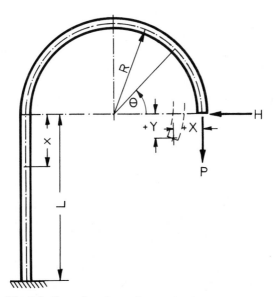

Fig. 8-8. Curved-end cantilever with half-circle bend

Substituting Eqs. (168), (169), (172) and (173) into Eq. (170), and, ignoring for the time being the terms in the bending moment equations involving H, we obtain, after integration, the deflection of the curved-end cantilever due to load P alone.

$$Y = \frac{PR^3}{EI} (0.3333 \, k^3 + k^2 + k + 0.3562) \tag{174}$$

Taking, in turn, the partial derivatives of the bending moments Eqs. (168) and (169) with respect to horizontal load H, gives

$$\frac{\partial M_1}{\partial H} = -R \sin \theta \tag{175}$$

$$\frac{\partial M_2}{\partial H} = -R \tag{176}$$

Since we are interested here in the horizontal deflection due to load P only, horizontal load H may be regarded in this part of the derivation as a fictitious quantity introduced only for the purpose of establishing the appropriate partial derivatives, Eqs. (175) and (176). Hence substituting Eqs. (175) and (176) together with Eqs. (168) and (169) into the Castigliano equation (171) integrating and cancelling the terms involving H, yields

$$X = \frac{PR^3}{EI} (0.5k^2 + k + 0.5) \tag{177}$$

The design formulas given by Eqs. (174) and (177) have been developed on the assumption that $H = 0$. By reversing the role of P and H next, and following identical procedure with that of the previous case, the deflection formulas due to load H are obtained

$$Y = \frac{HR^3}{EI} (0.5k^2 + k + 0.5) \tag{178}$$

$$X = \frac{HR^3}{EI} (k + 0.7854) \tag{179}$$

It may be observed here again that Eq. (178) can be obtained directly from Eq. (177) because Maxwell's theorem of reciprocal deflections applies. The resultant vertical and horizontal displacements of the free end of the curved member shown in Fig. 8-7 for the combined action of P and H can be obtained by the method of superposition. For the case illustrated in Fig. 8-7 the vertical and horizontal component deflections are directly additive. The bending moments at station B and C may be found by substituting $\theta = \pi/2$ in Eq. (168) and then $x = L$ in Eq. (169). The maximum bending moment found at station C is equal to

$$M = -R\left[H + P\left(1 + k\right)\right] \tag{180}$$

The interpretation of the resultant stresses and deflections must be, of course, made with due regard to the initial sign convention.

Curved-end Cantilever with Half-Circle Bend

The bending moments and deflections for the case illustrated in Fig. 8-8 are considered next. For the curved portion the general bending moment equation is given by Eq. (168) as before. When $\theta = \pi$, Eq. (168) yields

$$M_1 = -2PR$$

The bending moment along the straight portion of the beam is then

$$M_2 = Hx - 2PR \tag{181}$$

Taking separately the component deflections due to P and H, for the case given in Fig. 8-8 produces the following results:

$$Y = \frac{PR^3}{EI}\left(4k + 4.7124\right) \tag{182}$$

$$X = \frac{PR^3}{EI}\left(2 - k^2\right) \tag{183}$$

$$Y = \frac{HR^3}{EI}\left(2 - k^2\right) \tag{184}$$

$$X = \frac{HR^3}{EI}\left(0.3333k^3 + 1.5708\right) \tag{185}$$

Note that for relatively small values of k, all the above equations yield positive values in accordance with the convention indicated in Fig. 8-8. However when $k > \sqrt{2}$, Eqs. (183) and (184) give negative results. Hence in these cases horizontal deflection due to P and vertical deflection due to H must be opposite to those shown in Fig. 8-8.

Numerical Examples

Design Problem 9: Arched cantilever beam, shown in Fig. 8-9 is subjected to a concentrated end load W, inclined 45 deg to the horizontal axis. If the curved portion subtends the angle $\phi = 45$ deg and the length to radius ratio $k = 1$, derive the formulas for the maximum

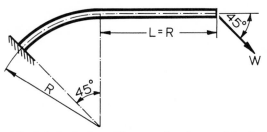

Fig. 8-9. Arched cantilever under skew end load

bending stress and the resultant deflection. Assume rectangular cross-section of the beam with b and h as width and depth, respectively.

Solution: The load components in the vertical and horizontal sense equivalent to skew load W are

$$P = H = \sqrt{2}\,W/2$$

The maximum bending moment follows from Eq. (143) where $M_o = 0$, $\theta = 45$ deg and $k = 1$ are substituted. This gives

$$M = -\sqrt{2}\,WR$$

and the corresponding bending stress is

$$S_b = \frac{M}{Z} = -\frac{\sqrt{2}\,WR}{bh^2/6} = -8.48\,\frac{WR}{bh^2}$$

The total vertical displacement is obtained by the algebraic addition of the deflections calculated from formulas Eqs. (149) and (162)

$$Y = \frac{WR^3}{bh^3E}\left(\frac{68\sqrt{2} - 96 + 15\pi\sqrt{2}}{4}\right) = 16.66\,\frac{WR^3}{bh^3E}$$

The total horizontal displacement follows from Eqs. (146) and (163) by substitution of the specified data and algebraic addition.

$$X = \frac{WR^3}{bh^3E}\left(\frac{15\pi\sqrt{2} - 96 + 24\sqrt{2}}{4}\right) = 1.14\,\frac{WR^3}{bh^3E}$$

Therefore the resultant displacement of the free end of the cantilever is obtained by vectoral addition.

$$u = \frac{WR^3}{bh^3E}\sqrt{16.66^2 + 1.14^2} = \frac{100\,WR^3}{6bh^3E}$$

If the absolute value of the maximum bending stress is taken as a criterion the above resultant displacement may be expressed as a function of stress, giving the following useful formula:

$$u = \frac{2R^2 S_b}{hE}$$

In the above numerical work a slide rule was used and several numbers were rounded off. The accuracy of the formulas derived in this way will be sufficient for most practical purposes.

Design Problem 10: A tubular steel support member with a 3-in. outer diameter and thickness of $t = 0.25$ in. is made to the shape and dimensions shown in Fig. 8-10 and carries a concentrated vertical load $P = 200$ lb. Assume that this member is fixed rigidly at ground level.

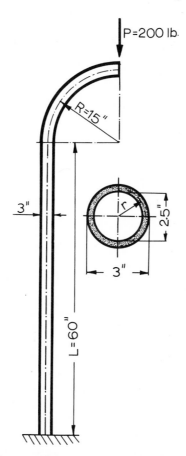

Fig. 8-10. Tubular curved-end cantilever

Calculate the magnitude and direction of end displacement if the modulus of elasticity $E = 28 \times 10^6$ psi, ignoring the effect of tube flattening due to bending.

Solution: Displacement in the direction of load is given by formula, Eq. (179)

$$X = \frac{PR^3}{EI} (k + 0.7854)$$

Mean radius of the tubular cross-section is $r = 1.375$ in. Straight length to radius ratio $k = 60/15 = 4$. Utilizing the approximate formula for the moment of inertia and substituting the relevant numerical values, the above equation yields

$$X = \frac{PR^3}{E\pi r^3 t} (k + 0.7854)$$

$$X = \frac{200 \times 15^3 (4 + 0.7854)}{28 \times 10^6 \times \pi \times 1.375^3 \times 0.25} = 0.057 \text{ in.}$$

Displacement perpendicular to the direction of load application is found with the aid of formula, Eq. (178)

$$Y = \frac{PR^3}{EI} (0.5k^2 + 0.5 + k)$$

Hence

$$Y = 0.057 \frac{(0.5k^2 + 0.5 + k)}{(k + 0.7854)} = 0.057 \times 12.5/4.8 = 0.148 \text{ in.}$$

The resultant displacement

$$u = \sqrt{X^2 + Y^2} = \sqrt{0.057^2 + 0.148^2} = 0.159 \text{ in.}$$

The direction of the resultant displacement is given by the tangent of the angle of inclination,

$$\frac{X}{Y} = 0.057/0.148 = 0.385$$

The angle corresponding to this tangent is approximately 21 deg.

Design Problem 11: Curved-end cantilever with half-circle bend illustrated in Fig. 8-8 is subjected to a system of forces H and P. Find the ratio H/P such as to make vertical component of deflection due to H and the bending moment due to the combined effect of H and P equal to zero.

Solution: The vertical displacement due to horizontal component

H is given by formula, Eq. (184). For the condition of zero vertical displacement

$$\frac{HR^3}{EI}\,(2 - k^2) = 0$$

from which $k = \sqrt{2}$

The bending moment at the base of the curved-end cantilever follows from Eq. (181)

$$M = R\,(Hk - 2P)$$

Making this moment vanish at the built-in end, and introducing $k = \sqrt{2}$, gives the required ratio

$$H/P = \sqrt{2}$$

Design Problem 12: A support bracket made of aluminum in form of an arched cantilever, subtending 270 deg and dimensioned as shown in Fig. 8-11, deflects vertically 0.2 inches under the working load P.

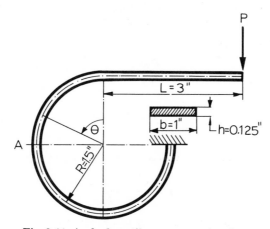

Fig. 8-11. Arched cantilever support bracket

Calculate the maximum bending stress compatible with the specified deflection. Take $E = 10 \times 10^6$ psi.

Solution: Since $k = 3/1.5 = 2$ and $\phi = 270$ deg, as shown in the sketch, Eq. (149) becomes

$$Y = \frac{PR^3}{EI}\left(\frac{80 + 81\pi}{12}\right)$$

if

$$I = bh^3/12 \text{ and } b = 1$$

$$Y = \left(\frac{P}{E}\right)\left(\frac{R}{h}\right)^3 (80 + 81\pi) = 334.34 \left(\frac{P}{E}\right)\left(\frac{R}{h}\right)^3$$

From Fig. 8-11, $R/h = 1.5/0.125 = 12$

Therefore solving for P gives

$$P = \frac{YE}{334.34 \times 1728} = \frac{0.2 \times 10 \times 10^6}{334.34 \times 1728} = 3.5 \text{ lb}$$

The maximum bending moment for this structure is found at A equal to $M = -3PR$. Hence the corresponding bending stress

$$S_b = -\frac{M}{Z} = -\frac{3PR}{bh^2/6} = -\frac{18PR}{bh^2}$$

Substituting the relevant numerical values gives

$$S_b = \frac{-18 \times 3.5 \times 1.5}{0.0156} = 6058 \text{ psi}$$

Deflection formula, Eq. (149) may be combined with the above stress formula in the following way. Let $F(k, \phi)$ represent a geometrical parameter. Then we have

$$Y = \frac{PR^3}{EI} F(k, \phi)$$

Since,

$$S_b = \frac{Mh}{2I} = \frac{3PRh}{2I}$$

Eliminating P between the stress and deflection formulas gives

$$S_b = \frac{3 \, YEh}{2R^2 F(k, \phi)}$$

This formula applies to curved cantilevers for which $\phi \geq \pi/2$.

Symbols for Chapter 8

b	Width of rectangular section, in.
E	Modulus of elasticity, psi
$F(k, \phi)$	Function of k and ϕ, (Eq. 149)
H	Horizontal load, lb
h	Depth of cross-section, in.

I	Moment of inertia, in.^{4}
$k = L/R$	Straight length to radius ratio
L	Length of straight portion, in.
L_e	Equivalent length, in.
M	Bending moment, lb-in.
M_o	Externally applied bending couple, lb-in.
M_1, M_2, M_B	Bending moments for various portions, lb-in.
P	Vertical load, lb
R	Mean radius of curvature, in.
r	Mean radius of tube, in.
S_b	Bending stress, psi
t	Wall thickness, in.
u	Resultant deflection, in.
W	Skew load, lb
X	Horizontal deflection, in.
x	Arbitrary distance, in.
Y	Vertical deflection, in.
Y_c	Maximum deflection of cantilever, in.
Z	Section modulus, in.^{3}
θ	Angle at which forces are considered, rad
ϕ	Angle subtended by curved member, rad
ψ	Slope, rad

CHAPTER NINE

Complex Flat Springs*

Assumptions

In the design of certain types of mechanical springs the engineer frequently deals with a complex-shaped flat spring with various combinations of straight and curved portions. A major portion of this chapter is devoted to detailed methods of analysis and the development of simple design equations which can be applied to the variety of complex-shaped flat springs, (Refs. 23 and 25).

It will be assumed that all forces applied to a spring are steady and are delivered to the structure without shock. The cross-sectional areas of spring elements are constant and have an axis of symmetry. The position of the neutral surface of the member coincides with the central surface. The elastic modulus is the same for tension as well as compression. The theory applies to small deflections only and the method of superposition is valid because of the linearity of the governing relations between the load and deflection. This method is very useful in this case because it permits to reduce a complex load and support condition into a combination of simpler conditions.

Three-quarter Circular Spring

In various applications a precurved cantilever type leaf spring can

---

* Portions of this chapter have been reprinted with permission from the copyrighted article "Deflections and Stresses of Complex Flat Springs" which appeared in the October 2, 1961 issue of PRODUCT ENGINEERING.

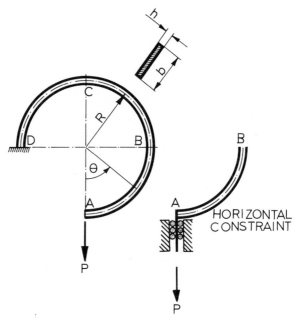

Fig. 9-1. Three-quarter circular spring

be employed as a loading or supporting member. This type of spring is usually fixed at one end while the other carries a concentrated load acting in the plane of curvature. The design problem consists usually of the calculation of stresses and deflections. However, many machine elements such as leaf springs are designed primarily on the basis of their elastic deflections. In this study therefore, special attention is paid to deflection although stress formulas are also given.

Let us now consider the first example of a complex spring in this category. The shape of this spring as well as the manner of loading are shown in Fig. 9-1. The bending moment at any point on the circular spring in terms of θ measured from line AC is

$$M = P R \sin \theta \, * \tag{186}$$

It follows directly from Eq. (186) that we have zero bending moment and zero bending stress at A and C where $\sin \theta$ is zero. Also the two regions of the maximum bending moment are at B and D where $\sin \theta = 1$. Consequently, the maximum bending stress can be obtained

---

* For meaning of symbols and dimensional units involved for this and other equations in this chapter see material at end of chapter.

directly from the well-known relation of strength of materials, $S_b = M/Z$. Hence we have

$$S_b = 6PR/bh^2$$

Shear stresses are present at the points of zero bending moment but their contribution to the stresses and strains in this particular case is negligible because the spring is sufficiently slender.

Derivation of the equation for the deflection under the spring load P can be again accomplished by using Castigliano's theorem. Since we are dealing here with so called slender members, it is only necessary to take into account the elastic strain energy due to bending. The statement of Castigliano's principle in polar coordinates applicable to this case is given by Eq. (75). Hence differentiating Eq. (186) with respect to P and utilizing Eq. (75), gives

$$Y = \frac{3\pi PR^3}{4EI} \tag{187}$$

The above formula gives the vertical displacement under load P. Suppose now that the free end of the leaf spring shown in Fig. 9-1 is constrained by guides to move in a vertical direction only. Then the above analysis should be modified by introducing a constraining couple M_f as a redundant quantity and the modified bending moment, Eq. (186), becomes

$$M = PR \sin \theta - M_f \tag{188}$$

Due to the constraint, the angle of rotation of the guided end is zero. According to the second principle of Castigliano, referred to in Chapter 2, the following equation can be used in calculating M_f

$$\int_0^{3\pi/2} M \frac{\partial M}{\partial M_f}\, d\theta = 0 \tag{189}$$

From Eq. (188), $\partial M/\partial M_f = -1$. Hence substituting this value together with Eq. (188) into Eq. (189), gives

$$\int_0^{3\pi/2} (M_f - PR \sin \theta)\, d\theta = 0 \tag{190}$$

Solving Eq. (190), gives $M_f = 2PR/3\pi$, and Eq. (188) becomes

$$M = PR \left(\sin \theta - \frac{2}{3\pi} \right) \tag{191}$$

On substituting Eq. (191) together with the partial derivative $\partial M/\partial P$ from this moment equation into Eq. (75), and integrating Eq. (75) between the limits of 0 and $3\pi/2$, yields

$$Y = \frac{PR^3}{EI} \left(\frac{9\pi^2 - 8}{12\pi} \right) \tag{192}$$

Comparing Eqs. (192) and (187) indicates that the effect of the guide-constraint in this particular case is relatively small. However, this effect changes with the angle of spring-arc, and if desired, may be investigated with the aid of the above equations.

Snap-ring Spring

As the next example consider a snap-ring type leaf spring illustrated in Fig. 9-2. Due to the symmetry it is sufficient to analyze one half of

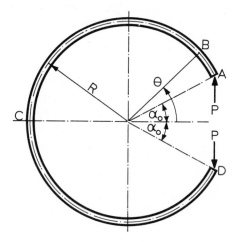

Fig. 9-2. Snap-ring spring

the spring such as AC. The bending moment equation set up with reference to point B is

$$M = PR \left(\cos \alpha_o - \cos \theta \right) \tag{193}$$

Hence for a given value of α_o, Fig. 9-2, the expression for the maximum bending stress becomes

$$S_b = \frac{6PR \left(1 + \cos \alpha_o \right)}{bh^2} \tag{194}$$

The total amount of deflection (change in distance AD) caused by the loads acting at the free ends equals twice the deflection of one half of the spring. Introducing the expression for bending moment, Eq. (193), and its partial derivative with respect to P into Eq. (75) and extending the integration between the limits of α_o and π results in the following deflection formula:

$$Y = \frac{PR^3}{EI}[(\pi - \alpha_o)\,(1 + 2\cos^2\alpha_o) + 1.5\sin 2\alpha_o] \qquad (195)$$

Here Y denotes the total change in distance AD caused by forces P. This and other equations are, of course, applicable when the direction of loading is reversed.

Circular Wave Spring

When the shape complexity of the leaf spring is progressively increased similar analyses to those of the two previous cases can be made without difficulty. Consider a leaf-type spring formed in a three-quarter circular wave, fixed at one end and subjected to a concentrated vertical load P at the other, Fig. 9-3. Portions AC and CF are analyzed separately and the corresponding bending moments are

$$\text{Part } AC \quad M = PR \sin \theta \qquad (196)$$
$$\text{Part } CF \quad M = PR\,(2 - \cos \theta) \qquad (197)$$

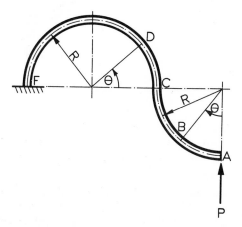

Fig. 9-3. Three-quarter circular wave spring

The maximum bending stress occurs at fixed end F and is equal to

$$S_b = \frac{18PR}{bh^2} \qquad (198)$$

The total resilience of the spring is the algebraic sum of the two component values of resilience corresponding to the portions AC and CF.

Therefore employing Eqs. (75), (196) and (197) gives

$$Y = \frac{PR^3}{EI} \int_0^{\pi/2} \sin^2 \theta \, d\theta + \frac{PR^3}{EI} \int_0^{\pi} (2 - \cos \theta)^2 \, d\theta$$

Integrating this equation gives the design formula for the deflection under load P

$$Y = \frac{19\pi PR^3}{4EI} \tag{199}$$

Adding a straight portion of length L, Fig. 9-4, does not influence the formula for the maximum bending stress, Eq. (198). The reason for

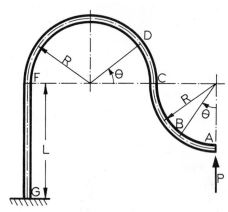

Fig. 9-4. Circular-wave spring with extension

this is that the extension FG acts as a cantilever beam, built-in at G and loaded by a couple equal to $3PR$ at F. The effect of direct stress caused by load P can be safely ignored because bending stresses in a slender member predominate.

The magnitude of deflection for the modified spring should however be corrected by considering the additional resilience of cantilever FG, subjected to an end couple. When this is done the deflection formula becomes

$$Y = \frac{PR^3}{EI} \left(9k + \frac{19\pi}{4} \right) \tag{200}$$

Design of U-Spring with Extension

Flat springs of a cantilever type, consisting of precurved and straight portions, are often employed as retaining brackets, holding clamps, loading members and other similar components. The curved portions in items are usually approximated by circular arcs; therefore, design formulas developed on the assumption of circular curvature become sufficiently accurate for most practical purposes.

The stress and deflection study of complex-shaped springs and similar structural members should first include the sketch of the bending moments. The investigation of the bending moment distribution along the contour of the spring helps locate the regions of zero bending moments or zero slopes so that the structure can be subdivided into simpler components for the purpose of analyzing the deflection. The total deflection can then be obtained by the well known method of superposition provided the elastic limit of the material under consideration is not exceeded.

Consider the U-spring with extension illustrated in Fig. 9-5. The distribution of the bending moment is shown in agreement with previous convention. The displacement of the loaded end C can be obtained by the method of superposition of the component deflections. The deflection of a symmetrical U-shape spring BC is

$$Y_{BC} = \frac{PR^3}{EI}\left(\frac{2k^3}{3} + \pi k^2 + 4k + \frac{\pi}{2}\right) \tag{201}$$

where $k = L_1/R$

The deflection of a straight cantilever beam AB of length $(L_2 - L_1)$ is given by a standard formula

$$Y_{AB} = \frac{P\,(L_2 - L_1)^3}{3EI} \tag{202}$$

Employing the method of superposition, the total deflection under the transverse load, $Y = Y_{AB} + Y_{BC}$. This gives

$$Y = \frac{PR^3}{EI}\left[\frac{n^3 + k^3}{3} + k^2\,(n + \pi) + k\,(4 - n^2) + \frac{\pi}{2}\right] \tag{203}$$

Here

$$n = \frac{L_2}{R},\, k = \frac{L_1}{R}$$

Maximum bending stress in this type of spring is either at A or D, depending on spring proportions. Therefore

When $(L_2 - L_1) > (L_1 + R)$

$$S_b = \frac{6P\,(L_2 - L_1)}{bh^2} = \frac{6PR}{bh^2}\,(n - k) \tag{204}$$

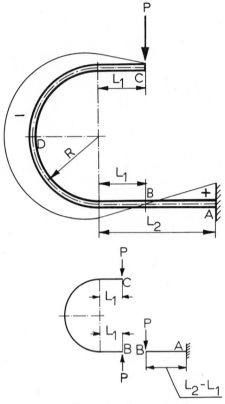

Fig. 9-5. U-spring, short extension

When $(L_2 - L_1) < (L_1 + R)$

$$S_b = \frac{6P\,(L_1 + R)}{bh^2} = \frac{6PR}{bh^2}\,(1 + k) \qquad (205)$$

When $L_2 = L_1$ a symmetrical U-shape spring is obtained whose deflection and maximum bending stress can be computed from Eqs. (201) and (205), respectively. However when $L_1 > L_2$ in Fig. 9-6, the spring cannot be divided into symmetrical components and therefore it is necessary to analyze this structure in three separate stages as follows:

Part AB $M_1 = Px_1$

Part BC $M_\theta = P\,(L_1 + R \sin \theta)$

Part CD $M_2 = P\,(L_1 - x_2)$

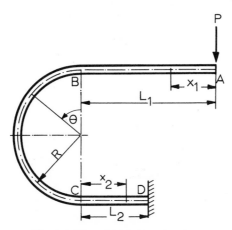

Fig. 9-6. U-spring, long extension

Hence the total deflection under load P, acting on a spring illustrated in Fig. 9-6, can be derived from the following expression

$$EI\,Y = \int_0^{L_1} M_1 \frac{\partial M_1}{\partial P}\,dx_1 + \int_0^{\pi} M_\theta \frac{\partial M_\theta}{\partial P}\,R\,d\theta + \int_0^{L_2} M_2 \frac{\partial M_2}{\partial P}\,dx_2$$

(206)

Integration of this equation results in the equation for deflection (in this particular case) identical with formula, Eq. (203). The maximum bending stress for this type of spring is always given by Eq. (205) provided the condition $L_1 \geqq L_2$ exists. The derivation methods discussed in relation to the springs in Figs. 9-5 and 9-6 can be extended to a wide variety of complex-shaped springs.

The procedure shows the importance of the preliminary analysis of the distribution of bending moment and the convenience of utilizing the Castigliano principle coupled with the method of superposition.

Closed U-spring

The theory developed previously for arched cantilevers can be applied directly to the closed U-spring shown in Fig. 9-7. However for such an equation as Eq. (149) to be applicable, the modifications which follow must be made. Let the component force P_1 act perpendicular to the straight portion L as shown in Fig. 9-7. Then

$$P_1 = P \cos v$$

(207)

If Y_1 denotes displacement of loaded end A in the direction of P_1 and

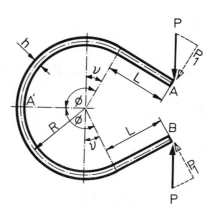

Fig. 9-7. Closed U-spring

Y the total change in the distance AB resulting from P then

$$Y_1 = 0.5Y \cos v \qquad (208)$$

The deflection caused by load P_1 is obtained directly from Eq. (149). Introducing the modified notations gives

$$Y_1 = \frac{P_1 R^3}{EI} F\,(k, \phi)$$

Substituting Eqs. (207) and (208) into the above equation results in the final deflection formula for the closed U-spring

$$Y = \frac{2PR^3\, F\,(k, \phi)}{EI} \qquad (209)$$

Thus in this particular case the deflection under the load P is independent of v. However when the maximum bending stress is considered, v appears in the relevant design formulas. For $\phi = \dfrac{\pi}{2} + v$ the expression for the maximum bending stress at A' becomes

$$S_b = \frac{6PR}{bh^2}\left(1 + \sin v + \frac{k}{\cos v}\right) \qquad (210)$$

When $v = 0$, Eq. (210) reduces to Eq. (205). On the other hand when $\phi \leqq \pi/2$, the stress is given by the equation

$$S_b = \frac{6PR}{bh^2}\,(1 - \cos \phi + k \cos v) \qquad (211)$$

When $v = 0$ and $\phi = \pi/2$ then Eq. (211) reduces to Eq. (205) also.

If desired, Eqs. (210) and (211) can be easily rearranged through the following substitutions: $v = \phi - \pi/2$ and $v = \pi/2 - \phi$, respectively. The spring proportions limit v to a value obtained from the geometrical relation, $k = \cot v$, provided load P is zero.

For a specified deflection Y and $\phi = \dfrac{\pi}{2} + v$, the limiting value becomes

$$Y = 2R \,(\cos v - k \sin v) \tag{212}$$

Equating formulas (209) and (212) gives

$$P = \frac{EI}{F\,(k,\phi)\,R^2}\,(\cos v - k \sin v) \tag{213}$$

This equation gives the value of P at which the deflected ends will touch without exerting any pressure. This condition can be accepted only if the corresponding bending stress, given by Eq. (210), does not exceed the elastic limit of the material.

The total amount by which the free ends of the S-spring in Fig. 9-8 approach each other under load P follows directly from Eq. (209).

$$Y = \frac{4PR^3}{EI}\,F\,(k,\phi) \tag{214}$$

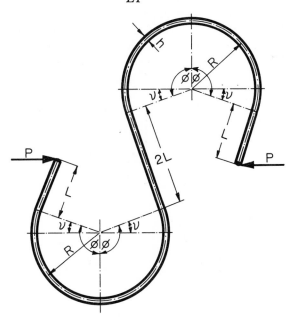

Fig. 9-8. S-spring

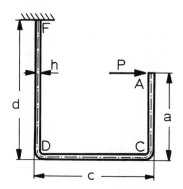

Fig. 9-9. Sharp-bend cantilever spring

The maximum bending stresses can be computed from Eqs. (210) and (211). Equations (212) and (213) are also applicable.

Sharp-bend Springs

Where springs involve relatively long straight portions and sharp radii of curvature, such as illustrated in Figs. 9-9 and 9-10, the preceding analysis must be applied cautiously. Sharp radii cause stress concentrations. There exists a considerable amount of design data on this particular aspect of structural behavior. As noted in Chapter 3 stress raisers are characteristic of elastic behavior. Plastic yielding reduces the effects of stress concentration, so that the practical sig-

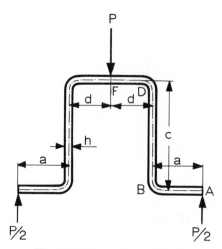

Fig. 9-10. Frame type spring

nificance of this problem depends on circumstances. Ductile material under static loading is usually not sensitive to stress raisers. The conditions involving fatigue also have considerable influence on breaking strength in presence of various types of stress raisers, and even ductility can prove to be a poor measure of immunity to stress concentration.

In the springs of Figs. 9-9 and 9-10 rather rapid stress variation is brought about by a curved beam effect of relatively small radius at the corners. The unit stresses in a beam of this type are not directly proportional to the distance of the neutral axis because the length of the elementary filaments is not identical. As a practical guide, as long as the ratio of depth of section to inner radius of the fillet is not more than about 0.5 the approximate method of stress calculation should yield acceptable results for most design purposes. Also, in calculating the deflections, the ratios of straight length to radius of curvature are usually quite large so that the effect of corner radii can be safely ignored. The cantilever spring in Fig. 9-9 can be subdivided into three portions, a, c and d. The corresponding bending moment equations expressed in terms of load P are as follows:

$$M_a = - Px \tag{215}$$

$$M_c = Pa \tag{216}$$

$$M_d = P (a - x) \tag{217}$$

The total displacement of the loaded end of the spring in the direction of P is

$$EI\,Y = \int_0^a M_a\,\frac{\partial M_a}{\partial P}\,dx + \int_0^c M_c\,\frac{\partial M_c}{\partial P}\,dx + \int_0^d M_d\,\frac{\partial M_d}{\partial P}\,dx \tag{218}$$

This equation illustrates the general procedure to be followed when the complex-shaped spring can be subdivided into a number of regions to which particular bending moment relations apply. Because a, c and d can be varied arbitrarily in this case the position and the magnitude of the maximum bending stress can be determined by inspection. Introducing Eqs. (215), (216) and (217) into Eq. (218), and integrating gives

$$Y = \frac{P}{3EI}\,(a^3 + 3a^2c + 3a^2d - 3ad^2 + d^3) \tag{219}$$

This equation is applicable also to such cases where $a = 0$, $a = c = 0$, $d = 0$ or $d = c = 0$ giving various design formulas for the calculation of the deflection.

When a frame-type leaf spring, Fig. 9-10, is loaded symmetrically, there must be a zero slope at F. This permits analysis of only one half of the structure, and the bending moments become

$$M_a = Px/2 \tag{220}$$

$$M_c = Pa/2 \tag{221}$$

$$M_d = P(a + x)/2 \tag{222}$$

The derivation can be carried out with the aid of Eq. (218) giving the following deflection formula:

$$Y = \frac{P}{6EI}\,[(a + d)^3 + 3a^2c] \tag{223}$$

In developing Eq. (223), it was assumed that the corner radii were relatively small and could be neglected. In more precise analysis location of the maximum bending stress in a frame-type spring should be determined with regard to possible stress concentrations at the corners. When the corner radii are sufficiently large to be neglected the maximum bending stress will occur at F under normal spring proportions. However, in addition to the above considerations the question of elastic stability of the vertical sides c may arise if the horizontal portions a and d are found to be relatively small.

Special Applications of U-Springs

In many design applications involving instruments and similar precision devices, a U-spring of a "double cantilever-cantilever" type shown in Fig. 9-11 is especially useful. This spring produces no torque along and perpendicular to the axis of displacement unlike most compression helical springs. This characteristic eliminates frictional forces at guides and pivots and the U-spring may be used successfully as a tension or compression device, (Ref. 26).

The derivation of the stress and deflection formulas for this type of spring may be accomplished either with the aid of Hooke's principle of elastic deformation or Castigliano theorem. The structure is basically statically indeterminate because the slope of the spring at the mount remains equal to zero whenever the spring is deflected. Because of the symmetry only one half of the spring need be analyzed and the bending moment equations for the straight and curved portions are

$$M_1 = M_f - Px \tag{224}$$

$$M_2 = M_f - PR\,(k + \sin\theta) \tag{225}$$

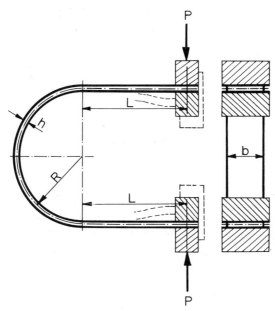

Fig. 9-11. U-spring with ends fixed as to slope

Here M_f is statically indeterminate and the equation for deriving its value is

$$\int_0^L M_1 \frac{\partial M_1}{\partial M_f} dx + \int_0^{\pi/2} M_2 \frac{\partial M_2}{\partial M_f} R\, d\theta = 0 \qquad (226)$$

Substituting $\dfrac{\partial M_1}{\partial M_f} = \dfrac{\partial M_2}{\partial M_f} = 1$, together with Eqs. (224) and (225) into Eq. (226), and integrating yields

$$M_f = PR \left(\frac{k^2 + \pi k + 2}{\pi + 2k} \right) \qquad (227)$$

Introducing $\phi = \pi/2$ in Eqs. (149) and (165), substituting the above expression for fixing moment M_f, and applying the method of superposition gives the total deflection of the U-spring

$$Y = \frac{PR^3}{EI} \left(\frac{0.33k^4 + 2.09k^3 + 4k^2 + 3.14k + 0.93}{2k + 3.14} \right) \qquad (228)$$

Numerical Examples

Design Problem 13: The complex-shaped flat spring made of spring steel, Fig. 9-12 is employed as a clip, going over a machine member

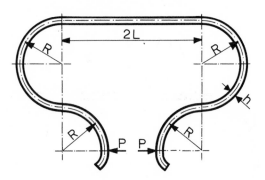

Fig. 9-12. Clip spring

by spreading apart the free ends. Calculate the maximum bending stress for a spring with the following requirements:

Mean radius of curvature	$R = 0.5$ in.
Straight portion	$2L = 1.5$ in.
Thickness of stock	$h = 0.040$ in.
Deflection of free ends	$2Y = 0.2$ in.
Modulus of elasticity	$E = 30 \times 10^6$ psi

Solution: The stress and deflection formulas applicable to this case are given by Eqs. (198) and (200), respectively. Since the load acting on the spring is not known, Eq. (198) is rearranged

$$P = \frac{S_b b h^2}{18R}$$

Introducing the above expression, and $I = bh^3/12$, into Eq. (200) gives the equation for maximum stress

$$S_b = \frac{6EhY}{R^2 (36k + 19\pi)}$$

By the definition, $k = L/R = 0.75/0.5 = 1.5$. Hence, substituting the numerical data, the maximum bending stress is

$$S_b = \frac{6 \times 30 \times 10^6 \times 0.040 \times 0.1}{0.25 (36 \times 1.5 + 19 \times 3.14)}$$

$$S_b = 25,300 \text{ psi}$$

Design Problem 14: A flat S-spring consists of two half-circles and carries a compression load P, Fig. 9-13. Calculate the required thickness of stock such that the maximum bending stress will not exceed

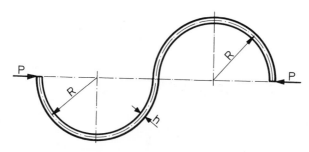

Fig. 9-13. Half-circle S-spring

30,000 psi when the spring is deflected 0.3 in. Also determine the resisting force at that deflection. Width of the material is 2 in., $R = 2.5$ in. and $E = 30 \times 10^6$ psi.

Solution: The deflection factor $F(k, \phi)$ from Eq. (149) and Eq. (211) together with Eq. (214) can be employed. Here, $\phi = \pi/2$ and $k = 0$ give $F(k, \phi) = \pi/4$. Hence Eq. (214) becomes

$$Y = \frac{12\pi PR^3}{bEh^3}$$

The maximum bending stress is found from Eq. (211) by substituting $k = 0$ and $\phi = \pi/2$. This gives

$$S_b = \frac{6PR}{bh^2}$$

Thus there are two equations and two unknowns. To solve, rearrange the above equation

$$P/h^2 = \frac{S_b b}{6R}$$

Substituting this expression in the deflection formula gives the required thickness of stock

$$h = \frac{2\pi R^2 S_b}{YE}$$

$$h = \frac{2 \times 3.14 \times 6.25 \times 30,000}{0.3 \times 30 \times 10^6}$$

$$h = 0.131 \text{ in.}$$

The corresponding compression load can be now found from the stress equation

$$P = \frac{bh^2 S_b}{6R} = \frac{30,000 \times 2 \times 0.131^2}{6 \times 2.5}$$

$$P = 69 \text{ lb}$$

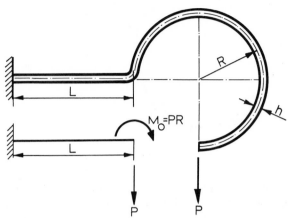

Fig. 9-14. Anchor spring

Design Problem 15: The anchor-type steel spring, Fig. 9-14, carries a load P at its free end. Calculate the downward displacement of the free end, and the maximum bending stress for the following dimensions:

Mean radius of curvature	$R = 1$ in.
Straight arm	$L = 2$ in.
Width of stock	$b = 0.5$ in.
Thickness of stock	$h = 0.1$ in.
Modulus of elasticity	$E = 30 \times 10^6$ psi
Concentrated load	$P = 15$ lb

Solution: Deflection of the curved portion alone is given by Eq. (187). The straight portion can be considered as a cantilever beam loaded with P and a bending couple $M_o = PR$, as illustrated. The slope at the end resulting from the couple is $M_o L/EI = PRL/EI$. The deflection resulting from M_o in the line of action of P is then

$$\frac{PRL}{EI} (L + R)$$

The component deflection because of the cantilever effect alone is $PL^3/3EI$. Hence adding the three components gives

$$Y = \frac{PR^3}{EI} \left(\frac{3\pi}{4} + \frac{k^3}{3} + k^2 + k \right)$$

Integration of Castigliano equations, will lead to the same result. The corresponding formula for the maximum bending stress follows directly from the geometry of the spring

$$S_b = \frac{6PR\,(1+k)}{bh^2}$$

Where $k = L/R = 2$. Substituting the numerical data gives

$$Y = \frac{15\,(2.356 + 2.666 + 4 + 2)\,12}{30 \times 10^6 \times 0.5 \times 10^{-3}} = 0.132\text{ in.}$$

$$S_b = \frac{6 \times 15 \times 3}{0.5 \times 0.01} = 54{,}000\text{ psi}$$

It is assumed that no severe stress raiser is present in this spring at the junction of the curved and straight portions.

Design Problem 16: A flat, mechanical spring is made of steel and formed in U-shape as shown in Fig. 9-15. If the modulus of elasticity

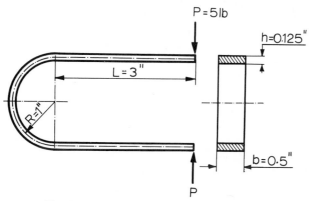

Fig. 9-15. Flat U-Spring without constraint

$E = 30 \times 10^6$ psi, calculate the decrease in distance between the free ends under concentrated end load $P = 5$ lb. The ratio of straight length L to the radius of curvature R is $k = 3$. The remaining dimensions are given in Fig. 9-15.

Solution: The expression for vertical deflection of one-half of the spring is given by Eq. (151). For $k = 3$ the approximate formula for this case becomes

$$Y = \frac{30PR^3}{EI} = \frac{360P}{bE}\left(\frac{R}{h}\right)^3$$

Substituting the known values for P, R, b, h and E gives

$$Y = \frac{360 \times 5 \times 8^3}{0.5 \times 30 \times 10^6} = 0.061 \text{ in.}$$

Since the structure is symmetrical, the total decrease in distance between the free ends is $2Y = 0.122$ in.

Design Problem 17: A double cantilever — cantilever type U-spring used in experimental analysis is loaded symmetrically as shown in Fig. 9-16. The material is aluminum with $E = 10 \times 10^6$ and the spring, made of flat stock, has the following nominal dimensions:

$$L = 3 \text{ in.}$$

$$R = 1 \text{ in.}$$

$$b = 0.25 \text{ in.}$$

$$h = 0.125 \text{ in.}$$

Assuming that the slope of the spring at the point of application of the external loading remains equal to zero for all elastic deflections calculate the theoretical spring rate of the device. Estimate the maximum bending stress and the corresponding total compression of the U-spring if the external load is $P = 10$ lb.

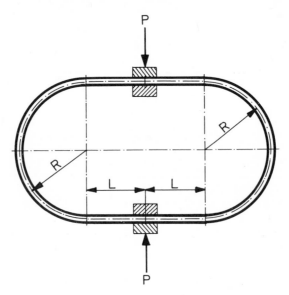

Fig. 9-16. Double U-spring with ends fixed as to slope

Solution: By definition, spring rate is obtained by dividing the load carried by the spring into the deflection produced by this load. Hence utilizing formula, Eq. (228) gives

$$\frac{P}{Y} = \frac{EI\,(4k + 2\pi)}{R^3\,(0.33k^4 + 2.09k^3 + 4k^2 + \pi k + 0.93)}$$

Since

$$I = \frac{bh^3}{12} = \frac{0.25 \times 0.125^3}{12} = 4.08 \times 10^{-5}\ \text{in.}^4$$

and $k = 3$, the required rate becomes

$$P/Y = 57\ \text{lb/in.}$$

Hence the deflection is

$$Y = \frac{P}{57} = \frac{10}{57} = 0.175\ \text{in.}$$

The maximum bending moment for this case follows from Eq. (227)

$$M_f = \frac{PR\,(2 + \pi k + k^2)}{2\pi + 4k}$$

$$M_f = \frac{10\,(2 + 3 \times 3.14 + 9)}{6.28 + 12} = 11.2\ \text{lb-in.}$$

and the corresponding bending stress is

$$S_b = \frac{M}{Z} = \frac{6M}{bh^2} = \frac{6 \times 11.2}{0.5 \times 0.125^2} = 8600\ \text{psi}$$

Symbols for Chapter 9

a	Length, in.
b	Width of rectangular section, in.
E	Modulus of elasticity, psi
$F\,(k, \phi)$	Function of k and ϕ, (Eq. 149)
h	Depth of cross-section, in.
I	Moment of inertia, in.^{4}
k	Straight length to radius ratio
L, L_1, L_2	Straight lengths, in.

M	Bending moment, lb-in.
M_f	Fixing moment, lb-in.
M_o	Externally applied bending couple, lb-in.
M_1, M_2, M_θ	Bending moments in U-springs, lb-in.
M_a, M_c, M_d	Bending moments in sharp-bend springs, lb-in.
n	Auxiliary length to radius ratio
P	Vertical load, lb
P_1	Component of vertical load, lb
R	Mean radius of curvature, in.
S_b	Bending stress, psi
x_1, x_2	Arbitrary distances, in.
Y	Vertical deflection, in.
Y_{AB}, Y_{BC}, Y_1	Components of vertical deflection, in.
Z	Section modulus, in.^{3}
α_o	Half-angle in snap ring, rad
θ	Angle at which forces are considered, rad
ϕ	Angle subtended by curved member, rad
ν	Auxiliary angle in U-springs, rad

CHAPTER TEN

Design of Thin Rings Loaded in Plane of Curvature

Assumptions

The analysis of thin circular rings is of importance in the design of machines and structures encountered in practically all branches of industry. The basic assumptions usually made in the calculations include the following:

1. Ring cross-section is uniform
2. Radial depth of cross-section is small in comparison with the mean radius of curvature. Ring is considered to be thin when the ratio of outer to inner ring diameter is not more than about 1.1
3. The strains are elastic
4. Deflections are small so that the ring does not lose the initially circular shape when loaded
5. External loads develop gradually so that no kinetic energy is supplied to the ring
6. The effect of strain energy caused by bending is predominant.

The continuous elastic ring is essentially a statically indeterminate structure. This means that the usual conditions of static equilibrium are insufficient to calculate the reactive forces and moments and it is necessary to supplement these conditions with the equations of elastic deformation. Hence in order to estimate the strength and rigidity characteristics of a closed ring the classical statically indeterminate problem must be solved first.

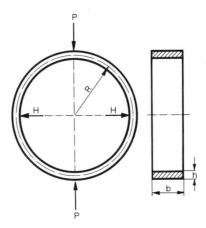

Fig. 10-1. Thin ring in compression

Thin Ring Theory

Consider a thin elastic ring subject to diametral tension and compression as shown in Fig. 10-1. In order to obtain the expression for the bending moment in terms of statically independent forces P and H, one quadrant, Fig. 10-2, may be singled out for the purpose of the discussion. For the convenience of the complete analysis, involving vertical and horizontal deflections, the ring is assumed to be loaded by a system of forces, P and H. Either of these forces can be regarded

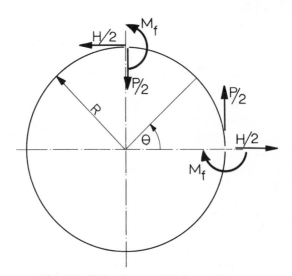

Fig. 10-2. Thin ring, equilibrium of forces

as real, or fictitious depending on specific requirements. For instance in considering the ring to be compressed along the vertical diameter the corresponding increase in the horizontal diameter can be calculated with the aid of the fictitious load method and the first theorem of Castigliano. In this case H is regarded as a fictitious quantity and P denotes real external loading.

One quadrant of the ring, shown in Fig. 10-2, remains in equilibrium under the system of forces $P/2$, $H/2$ and M_f. The vertical and horizontal load components follow from the conditions of symmetry and equilibrium. The fixing couple, M_f, is however statically indeterminate and must be found first before the relevant bending moment equation, at an arbitrary angle θ, can be defined in terms of compressive loading.

The general expression for the bending moment, involving real loading, is

$$M = \frac{PR}{2}\,(1 - \cos \theta) - M_f \,* \qquad (229)$$

Observing that there cannot be any rotation of the ring cross-section at the point of the application of the fixing couple, the second theorem of Castigliano states that the partial derivative of the total strain energy of the ring with respect to the fixing couple must be zero. Since in a thin ring the contribution of bending strain energy predominates, we can write

$$\frac{\partial U}{\partial M_f} = \frac{4R}{EI} \int_0^{\pi/2} M \frac{\partial M}{\partial M_f}\, d\theta = 0 \qquad (230)$$

From Eq. (229), $\partial M/\partial M_f = -1$. Hence, substituting for the partial derivative and the bending moment in Eq. (230), gives

$$M_f = \frac{\pi - 2}{2\pi}\, PR \qquad (231)$$

Introducing Eq. (231) into Eq. (229), yields

$$M = \frac{PR}{2\pi}\,(2 - \pi \cos \theta) \qquad (232)$$

The maximum bending moment is found from Eq. (232) when $\theta = \pi/2$ is substituted. This gives $M = PR/\pi$. Positive sign indicates here that the bending moment at the point of application of load P puts the outer surface of the ring in compression in accordance with previously established convention.

---

* For meaning of symbols and dimensional units involved for this and other equations in this chapter see material at end of chapter.

Once the bending moment is expressed in terms of statically independent force P, as shown by Eq. (232), the decrease in vertical diameter of the ring follows from Eq. (51). This yields

$$Y = \frac{4R}{EI} \int_0^{\pi/2} M \frac{\partial M}{\partial P} \, d\theta \tag{233}$$

Note that the above integration is, in effect, extended over the entire ring. Since from Eq. (232), $\partial M/\partial P = R \, (2 - \pi \cos \theta)/2\pi$, making the necessary substitutions and integrating Eq. (233), gives

$$Y = \frac{PR^3}{EI} \frac{(\pi^2 - 8)}{4\pi} \tag{234}$$

The above formula is well known to designers. The numerical results obtained from this formula are sufficiently accurate provided the mean radius of ring R, is large in comparison with the radial thickness of the ring. In order to determine the increase of the horizontal diameter, resulting from compression along the vertical diameter, use is made of the fictitious load method. The bending moment equation is now

$$M = \frac{PR}{2\pi} (2 - \pi \cos \theta) + \frac{HR}{2} \sin \theta \tag{235}$$

By analogy to Eq. (233), the horizontal deflection is

$$X = \frac{4R}{EI} \int_0^{\pi/2} M \frac{\partial M}{\partial H} \, d\theta \tag{236}$$

However, in this case, M is given by Eq. (235), instead of Eq. (232), and the relevant partial derivative, $\partial M/\partial H = \dfrac{R}{2} \sin \theta$. Solving Eq. (236) and remembering that the fictitious load H is finally made equal to zero, gives

$$X = \frac{PR^3}{EI} \frac{(4 - \pi)}{2\pi} \tag{237}$$

It may be of interest to note here that the ratio of vertical to horizontal deflection in this case is greater than one, and is equal to $Y/X = (\pi^2 - 8)/(8 - 2\pi) = 1.0893$. Also, making the bending given by Eq. (232), equal to zero, we find that the moment changes its sign at about $\theta = 50\frac{1}{2}$ deg.

Shear and Direct Stress Effects on Thin Rings

The question sometimes is asked what error is actually introduced in the analysis of thin rings by ignoring the effects of strain energy due to shear and direct compressive forces. To analyze this consider the general expression for the total elastic strain energy, Eq. (46). Since we are dealing with relatively thin rings the neutral axis can be assumed to coincide with the central axis and the shear stress may be taken to be uniformly distributed over the ring cross-section. Hence Eq. (46) can be simplified to read

$$U = \int_0^\theta \left[\frac{M^2}{2EI} + \frac{N^2}{2AE} + \frac{Q^2}{2AG} \right] R \, d\theta \qquad (238)$$

By analogy to Eq. (49), the relevant expressions for the deflections on vertical and horizontal diameters of the ring become

$$Y = 4R \int_0^{\pi/2} \left[\frac{M}{EI} \frac{\partial M}{\partial P} + \frac{N}{AE} \frac{\partial N}{\partial P} + \frac{Q}{AG} \frac{\partial Q}{\partial P} \right] d\theta \qquad (239)$$

and

$$X = 4R \int_0^{\pi/2} \left[\frac{M}{EI} \frac{\partial M}{\partial H} + \frac{N}{AE} \frac{\partial N}{\partial H} + \frac{Q}{AG} \frac{\partial Q}{\partial H} \right] d\theta \qquad (240)$$

In order to derive more exact formulas for deflections the normal and shear forces at an arbitrary section θ, are considered. The relevant expressions can include the real and fictitious quantities simultaneously. This gives

$$N = \frac{P \cos \theta}{2} - \frac{H \sin \theta}{2} \qquad (241)$$

and

$$Q = - \frac{P \sin \theta}{2} - \frac{H \cos \theta}{2} \qquad (242)$$

Utilizing Eq. (235), and Eqs. (239) through (242) the vertical and horizontal deflections become

$$Y = \frac{PR^3}{EI} \frac{(\pi^2 - 8)}{4\pi} + \frac{PR\pi}{4AE} + \frac{PR\pi}{4AG} \qquad (243)$$

and

$$X = \frac{PR^3}{EI} \frac{(4 - \pi)}{2\pi} - \frac{PR}{2AE} + \frac{PR}{2AG} \qquad (244)$$

In order to evaluate the contribution of bending, direct and shear stresses to deflection take for instance $G = 0.384E$, $A = bh$, $I = bh^3/12$ and the ratio $\chi = R/h$. This yields

$$Y = \frac{P\chi}{bE} \, (1.7856 \, \chi^2 + 0.7854 + 2.0453) \tag{245}$$

and

$$X = \frac{P\chi}{bE} \, (1.6392 \, \chi^2 - 0.5000 + 1.3020) \tag{246}$$

If the value of $\chi = 10$, is substituted in the above equations the percentage contribution of various stresses can be calculated directly. The results of these calculations are summarized in Table 10-1.

Table 10-1. Percentage Stress Contribution in Thin Rings

	Bending Stress	Direct Stress	Shear Stress
Vertical Deflection	+98.44	+0.43	+1.13
Horizontal Deflection	+99.51	−0.30	+0.79

Dividing Eq. (245) into Eq. (246), gives also

$$Y/X = \frac{1.7856 \, \chi^2 + 2.8307}{1.6392 \, \chi^2 + 0.8020}$$

It is evident that for a relatively large value of χ, the above expression gives 1.0893. For $\chi = 10$, we get $Y/X = 1.1012$.

The results illustrated in Table 10-1, indicate clearly that for relatively thin rings the effects of direct and shear stresses on deflections can be safely ignored. Such a simplification results in sizeable reduction of the amount of mathematical work normally associated with the development of working design formulas.

Ring Under Offset Loading

In performing the analysis of a thin circular ring subjected to a diametral compression it was admissible to analyze only one quarter of the ring because of complete symmetry of deformation about the two perpendicular axes. The case of a continuous circular ring carrying any two opposite and localized bending couples should be considered as having only one axis of symmetry. To illustrate an approach to the calculation of bending moment for such members consider a ring under diametrically opposed bending couples as shown in Fig. 10-3.

As in all the cases involving closed rings we have to deal here with a statically indeterminate member. In order to simplify the calculations the couples are assumed to act at the end of a diameter. However the reasoning developed in this special case applies also to all other cases involving opposite couples acting at the ends of a chord. Since the deformed shape of the ring, illustrated in Fig. 10-3, can be symmetrical only about the vertical diameter, the ring may be considered as consisting of two semi-circular members, fixed at the top and held in equilibrium by the horizontal forces H, fixing moment M_f

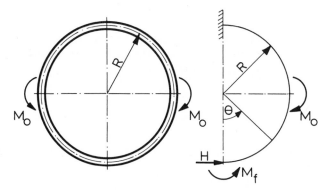

Fig. 10-3. Thin ring under opposed bending couples

and the externally applied bending couple M_o. By analogy to straight beams, known from the elementary strength of materials, we conclude that there must be two expressions for the bending moment applicable in the ranges $0 \leq \theta \leq \pi/2$ and $\pi/2 \leq \theta \leq \pi$. Denoting these bending moments by M_1 and M_2, respectively, gives

$$M_1 = HR (1 - \cos \theta) + M_f \qquad (247)$$

and

$$M_2 = HR (1 - \cos \theta) + M_f - M_o \qquad (248)$$

Since both H and M_f are considered here to be statically indeterminate quantities, we may call on the second theorem of Castigliano, expressing zero bending slope and zero tangential displacement at $\theta = 0$. The corresponding equations are

$$\int_0^{\pi/2} M_1 \frac{\partial M_1}{\partial H} \, d\theta + \int_{\pi/2}^{\pi} M_2 \frac{\partial M_2}{\partial H} \, d\theta = 0 \qquad (249)$$

$$\int_0^{\pi/2} M_1 \frac{\partial M_1}{\partial M_f}\, d\theta + \int_{\pi/2}^{\pi} M_2 \frac{\partial M_2}{\partial M_f}\, d\theta = 0 \qquad (250)$$

Note that in Eqs. (249) and (250) the terms referring to flexural rigidity and radius of the ring can be cancelled because they are simply multiplied by zero.

The required partial derivatives follow directly from Eqs. (247) and (248).

$$\frac{\partial M_1}{\partial H} = \frac{\partial M_2}{\partial H} = R\,(1 - \cos\theta) \qquad (251)$$

and

$$\frac{\partial M_1}{\partial M_f} = \frac{\partial M_2}{\partial M_f} = 1 \qquad (252)$$

Combining Eqs. (247) through (252), performing integration and substituting the relevant limits gives the redundant reactions, H and M_f, in terms of the applied external couple M_o.

$$H = 2M_o/\pi R$$

and

$$M_f = -\,(4 - \pi)\, M_o/2\pi$$

Substituting the above values into Eqs. (247) and (248), yields

$$M_1 = \left(\frac{1}{2} - \frac{2}{\pi}\cos\theta\right) M_o \qquad (253)$$

and

$$M_2 = -\left(\frac{1}{2} + \frac{2}{\pi}\cos\theta\right) M_o \qquad (254)$$

Equations (253) and (254) can be used in the calculation of bending strength of a thin ring subjected to diametrically opposed couples. The usual elementary formula, Eq. (8) can be employed here with success.

Ring with Horizontal Constraint

An interesting case of a horizontally constrained circular ring subjected to tensile loading is shown in Fig. 10-4. In order to evaluate the effect of constraint on the bending strength of such a curved member, consider one quadrant of the ring fixed at the top and held in equilibrium by a statically determinate reaction $P/2$, together with the two redundant reactions H and M_f. The thrust H is caused by

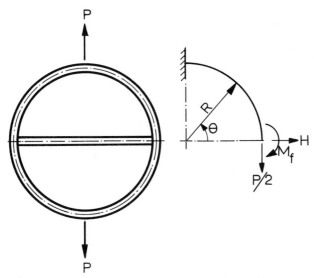

Fig. 10-4. Thin ring with horizontal constraint

the presence of the horizontal member which in this particular case will be assumed to be absolutely rigid.

In accordance with Fig. 10-4, and previously established sign convention, the bending moment at an arbitrary section defined by θ, is

$$M = HR \sin \theta - \frac{PR}{2} (1 - \cos \theta) - M_f \qquad (255)$$

Since the horizontal displacement and angular rotation of the ring, section adjacent to the rigid horizontal member, are zero the following Castigliano equations apply

$$\int_0^{\pi/2} M \frac{\partial M}{\partial H} \, d\theta = 0 \qquad (256)$$

and

$$\int_0^{\pi/2} M \frac{\partial M}{\partial M_f} \, d\theta = 0 \qquad (257)$$

From Eq. (255), the partial derivatives required for the solution of the above equations are

$$\frac{\partial M}{\partial H} = R \sin \theta$$

and

$$\frac{\partial M}{\partial M_f} = -1$$

Substituting the above derivatives and the bending moment, Eq. (255) into Eqs. (256) and (257), integrating and evaluating the limits of integration, yields the following simultaneous equations

$$\pi RH - 4M_f - PR = 0 \tag{258}$$

$$-4RH + 2\pi M_f + (\pi - 2)\, PR = 0 \tag{259}$$

Solving Eqs. (258 and (259) for H and M_f, gives

$$H = P\, \frac{(4 - \pi)}{(\pi^2 - 8)} \tag{260}$$

and

$$M_f = PR\, \frac{(4 + 2\pi - \pi^2)}{2\,(\pi^2 - 8)} \tag{261}$$

The above formulas for the redundant reactions, derived for the case illustrated in Fig. 10-4, are of special interest because their numerical values are the same as those which will be found later in the analysis of a built-in semicircular arch with a central concentrated load. It follows then that each half of the ring with a rigid horizontal constraining member will have the same stresses and deflections as those found in a semicircular arch with built-in supports.

Ring Under Radial Loading

The special case of a closed ring with symmetrical radial loading is shown in Fig. 10-5. Assume that the ring is held rigidly at the top bracket which resists two radial loads. If now the ring is cut at $\theta = 0$,

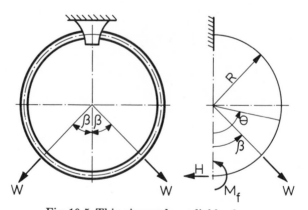

Fig. 10-5. Thin ring under radial loads

redundant forces H and M_f must be added at that section to assure equilibrium. Because of symmetry there is no rotation of the section at the bottom of the ring and no tangential displacement at the point of application of H. The redundant quantities H and M_f are unknown. In order to determine their values Castigliano's second theorem is applied as before. The bending moment expressions are as follows

$$M_1 = M_f - HR\ (1 - \cos \theta) \qquad (262)$$

and

$$M_2 = M_f - HR\ (1 - \cos \theta) + WR \sin\ (\theta - \beta) \qquad (263)$$

The Castigliano equations applicable to this case are essentially the same as Eqs. (249) and (250) with the exception of limits of integration which should now be 0 to β and β to π. The partial derivatives necessary for the solution of Eqs. (249) and (250) are essentially those given by Eqs. (251) and (252). Note however that the sign of $\partial M_1/\partial H$ and $\partial M_2/\partial H$ is opposite because of the difference in sign for H in Eqs. (247) and (262). Therefore the Castigliano equations written in full for this case give

$$\int_0^\beta [HR\ (1 - \cos \theta)^2 - M_f\ (1 - \cos \theta)]\ d\theta$$

$$+ \int_\beta^\pi [HR\ (1 - \cos \theta)^2 - M_f\ (1 - \cos \theta) - WR\ (\sin \theta \cos \beta$$

$$- \cos \theta \sin \beta)\ (1 - \cos \theta)]\ d\theta = 0 \qquad (264)$$

and

$$\int_0^\beta [M_f - HR\ (1 - \cos \theta)]\ d\theta + \int_\beta^\pi [M_f - HR\ (1 - \cos \theta)$$

$$+ WR\ (\sin \theta \cos \beta - \cos \theta \sin \beta)]\ d\theta = 0 \qquad (265)$$

Noting that some of the functions in the above equations remain unchanged for any value of θ, as evident from Fig. 10-5, equations (264) and (265) may be simplified by combining certain limits of integration. This yields

$$\int_0^\pi [HR\ (1 - \cos \theta)^2 - M_f\ (1 - \cos \theta)]\ d\theta -$$

$$\int_\beta^\pi [WR\ (\sin \theta \cos \beta - \cos \theta \sin \beta)\ (1 - \cos \theta)]\ d\theta = 0$$

$$(266)$$

and

$$\int_0^\pi [M_f - HR\,(1 - \cos\theta)]\,d\theta + \int_\beta^\pi [WR\,(\sin\theta\cos\beta - \cos\theta\sin\beta)]\,d\theta = 0$$

(267)

Integrating Eqs. (266) and (267) between the limits indicated and simplifying the trigonometric functions involved, gives

$$3\pi HR - 2\pi M_f - WR\,[2 + 2\cos\beta + (\pi - \beta)\sin\beta] = 0 \quad (268)$$

$$-\pi HR + \pi M_f + WR\,(1 + \cos\beta) = 0 \quad (269)$$

Hence, solving Eqs. (268) and (269) for H and M_f yields the desired redundant quantities

$$H = \frac{W\,(\pi - \beta)\sin\beta}{\pi} \quad (270)$$

and

$$M_f = \frac{WR}{\pi}\,[(\pi - \beta)\sin\beta - (1 + \cos\beta)] \quad (271)$$

Note that when $\beta = 0$, Eqs. (270) and (271) give $H = 0$ and $M_f = -2\,WR/\pi$. This value is essentially the same as that given by Eq. (232), for $\theta = \pi/2$, provided $W = P/2$.

Piston Type Rings

In some cases a piston type ring of uniform cross-section, Fig. 10-6, may be applied in machine design. If Y denotes the initial gap in the ring at q = 0, the load per unit length q, necessary to close the gap may be calculated in the following manner. Consider an auxiliary angle ε, smaller than θ. The elementary bending moment taken about a point defined by θ, is $dM_q = qRd\varepsilon \times R\sin(\theta - \varepsilon)$. Hence the total bending moment is

$$qR^2 \int_0^\theta \sin(\theta - \varepsilon)\,d\varepsilon = qR^2\,(1 - \cos\theta)$$

Introducing fictitious load P, the bending moment equation becomes

$$M = qR^2\,(1 - \cos\theta) + PR\,(1 - \cos\theta) \quad (272)$$

Castigliano principles applied to the model shown in Fig. 10-6, gives

$$Y = \frac{2R}{EI}\int_0^\pi M\,\frac{\partial M}{\partial P}\,d\theta \quad (273)$$

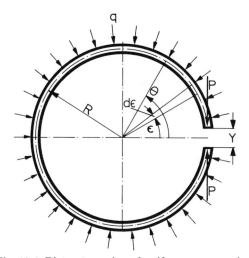

Fig. 10-6. Piston type ring of uniform cross-section

Substituting Eq. (272), and $\partial M/\partial P = R\,(1 - \cos\theta)$, into Eq. (273), and integrating between the limits indicated, yields

$$Y = \frac{3\pi q R^4}{EI} \qquad (274)$$

Since many rings in practice have rectangular cross-sections, substituting Eq. (7) for I, in Eq. (274), and solving for q, gives

$$q = \frac{YEbh^3}{36\pi R^4} \qquad (275)$$

When $\theta = \pi$, the maximum bending moment follows from Eq. (272), for $P = 0$. The corresponding bending stress is

$$S_b = \frac{12qR^2}{bh^2} \qquad (276)$$

In the above equations, q is expressed in pounds per inch of circumference. If this unit load is divided by the width of the ring cross-section, external pressure in pounds per square inch is obtained.

When the ring shown in Fig. 10-6, has varying cross-section the maximum bending stress at an arbitrary cross-section is

$$S_b = \frac{6\,qR^2\,(1 - \cos\theta)}{bh^2} \qquad (277)$$

At $\theta = \pi$, Eq. (277) reduces to Eq. (276). For a fixed value of the working stress S_b, external load q, and width of the ring b, Eq. (277) gives

$$h = 2.45\, R\, \sqrt{\frac{q}{b}\, \frac{(1 - \cos \theta)}{S_b}} \qquad (278)$$

It is therefore possible to select such values of ring thickness as a function of θ, for which the external pressure and the bending stress will remain constant. In the design of piston rings this will assure a uniform pressure loading against the inner wall of the cylinder.

Rotating Ring

When a closed circular ring is rotating in its plane of curvature the analysis of stresses may be made by considering the diagram shown in Fig. 10-7. At the horizontal diameter opposite and equal in magnitude internal forces P, hold the upper and lower portions of the ring together. If uniform radial load q, represents the centrifugal force per unit length of ring periphery, the force corresponding to the ring element $d\theta$, is $qRd\theta$. The horizontal component of this force is balanced by a similarly situated component on the opposite side of the y-y axis because of symmetry. However the sum of all vertical components produces tensile stresses at x-x sections of the ring. This sum is given by the following elementary equation of equilibrium:

$$2 \int_0^{\pi/2} qR \sin \theta\, d\theta = 2P$$

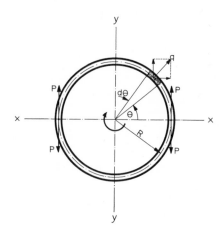

Fig. 10-7. Rotating ring in plane of curvature

which gives
$$P = qR \qquad (279)$$
The centrifugal load per unit length is
$$q = \frac{w}{g} \frac{V^2}{R} \qquad (280)$$
In the above equation, V denotes the peripheral velocity of the ring at the centerline, w is the weight of the ring per unit length and, g denotes acceleration due to gravity. The tensile stress in the ring is given by the usual elementary formula, $S = P/A$. Since the volume of the ring element of unit length is numerically equal to the cross-sectional area A, the specific weight of the ring material $\gamma = w/A$. Hence combining Eqs. (279) and (280) with the above stress formula gives

$$S = \frac{\gamma V^2}{g} \qquad (281)$$

This simple formula defines the tensile stress in a rotating ring. The specific weight of a metal γ, is often given in lb/in³. When this is the case it should be remembered to express V, in the above equation, in in/sec, and put $g = 386.4$ in/sec². When this is done Eq. (281), will yield stress in psi. The above quoted formulas apply to relatively thin rings and the tensile stress caused by the centrifugal loading is assumed to be uniformly distributed over the ring cross-section.

Simplified Tables and Charts for Circular Rings

The preceding discussion of thin elastic rings indicates that even in the elementary cases of loading and support the calculations may be time consuming. The solutions involve usually numerous trigonometric terms and the problem is complicated further by the static indeterminacy. A comprehensive study of circular rings has been made by Biezeno and Grammel (Ref. 27), to which the reader with broader theoretical interests is referred.

This section brings forth tables and charts for more common cases of circular rings loaded in plane of curvature. This information has been extracted from the published data (Ref. 22) and concerns radial deflection, bending slope and bending moment at an arbitrary ring cross-section. The type of loading analyzed includes concentrated external loads, external bending couples cosine load distribution and combinations thereof. Symbols u, ψ and M refer to deflection, slope, and bending moment, respectively. The corresponding design factors are denoted by K_u, K_ψ and K_M. The design equations are given in Table 10-2 through 10-11. These equations are illustrated graphically in Figs. 10-8 through 10-17.

Table 10-2. Equations for Circular Ring in Diametral Tension, Fig. 10-8

Symbol	Function	Range of Application
K_u	$0.2500 \sin \theta + (0.3927 - 0.2500\,\theta) \cos \theta - 0.3183$	$0 - \pi$
K_ψ	$(0.2500\,\theta - 0.3927) \sin \theta$	$0 - \pi$
K_M	$0.5000 \sin \theta - 0.3183$	$0 - \pi$

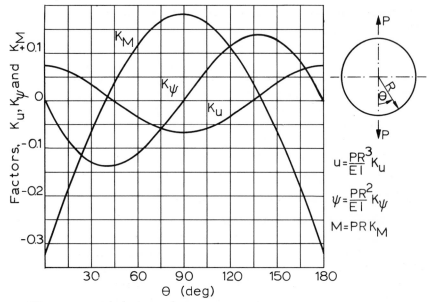

Fig. 10-8. Design factors for circular ring in diametral tension, Table 10-2

Table 10-3. Equations for Circular Ring in Four Way Tension, Fig. 10-9

Symbol	Function	Range of Application
K_u	$0.2500\,(1 + \theta) \sin \theta + (0.6427 - 0.2500\,\theta) \cos \theta - 0.6366$	$0 - \pi/2$
K_ψ	$0.2500\,(\theta - 1.5708) \sin \theta + 0.2500\,\theta \cos \theta$	$0 - \pi/2$
K_M	$0.5000\,(\sin \theta + \cos \theta) - 0.6366$	$0 - \pi/2$
	$0.500\,(\sin \theta - \cos \theta) - 0.6366$	$\pi/2 - \pi$

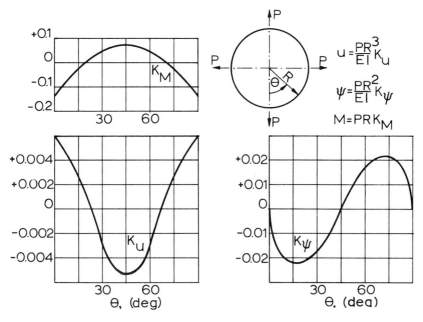

Fig. 10-9. Design factors for circular ring in four way tension, Table 10-3

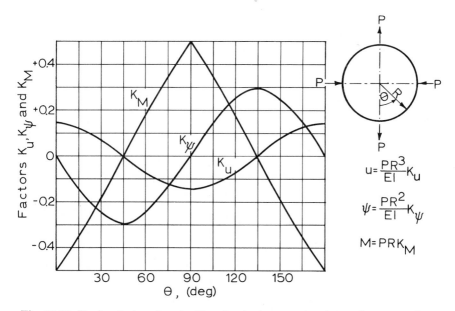

Fig. 10-10. Design factors for circular ring in two way tension and compression,
Table 10-4

Table 10-4. Equations for Circular Ring in Two Way Tension and
Compression, Fig. 10-10

Symbol	Function	Range of Application
K_u	$0.2500 \, (1 - \theta) \sin \theta + (0.1427 - 0.2500 \, \theta) \cos \theta$	$0 - \pi/2$
K_ψ	$0.2500 \, (\theta - 1.5708) \sin \theta - 0.2500 \, \theta \cos \theta$	$0 - \pi/2$
K_M	$0.5000 \, (\sin \theta - \cos \theta)$	$0 - \pi/2$
	$0.5000 \, (\sin \theta + \cos \theta)$	$\pi/2 - \pi$

Table 10-5. Equations for Circular Ring Under Parallel Forces, Fig. 10-11

Symbol	Function	Range of Application
K_u	$0.1427 \cos \theta - 0.0966$	$0 - \pi/4$
	$0.7500 \sin \theta + 0.5000 \, (1.5708 - \theta) \cos \theta - 0.8037$	$\pi/4 - 3\pi/4$
K_ψ	$- 0.1427 \sin \theta$	$0 - \pi/4$
	$0.5000 \, (\theta - 1.5708) \sin \theta + 0.2500 \cos \theta$	$\pi/4 - 3\pi/4$
K_M	$- 0.0966$	$0 - \pi/4$
	$\sin \theta - 0.8037$	$\pi/4 - 3\pi/4$
	$- 0.0966$	$3\pi/4 - \pi$

Table 10-6. Equations for Circular Ring Under Diametrically
Opposed Couples, Fig. 10-12

Symbol	Function	Range of Application
K_u	$0.5000 - 0.3183 \, \theta \sin \theta - 0.4775 \cos \theta$	$0 - \pi/2$
K_ψ	$0.1592 \sin \theta - 0.3183 \, \theta \cos \theta$	$0 - \pi/2$
K_M	$0.5000 - 0.6366 \cos \theta$	$0 - \pi/2$
	$-0.5000 - 0.6366 \cos \theta$	$\pi/2 - \pi$

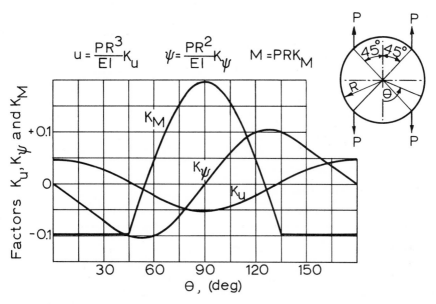

Fig. 10-11. Design factors for circular ring under four parallel forces, Table 10-5

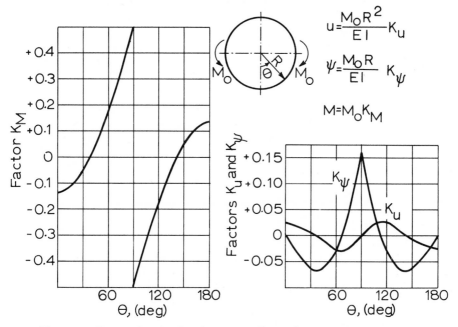

Fig. 10-12. Factors for circular ring under diametrically opposed couples,
Table 10-6

Table 10-7. Equations for Circular Ring Under Four Couples, Fig. 10-13

Symbol	Function	Range of Application
K_u	$1 - 0.4502\ \theta \sin\theta - 1.0288 \cos\theta$	$0 - \pi/4$
	$0.7071 \sin\theta - 0.4502\ \theta \sin\theta - 0.3217 \cos\theta$	$\pi/4 - 3\pi/4$
K_ψ	$0.5786 \sin\theta - 0.4502\ \theta \cos\theta$	$0 - \pi/4$
	$0.7071 \cos\theta - 0.1285 \sin\theta - 0.4502\ \theta \cos\theta$	$\pi/4 - 3\pi/4$
K_M	$1 - 0.9003 \cos\theta$	$0 - \pi/4$
	$- 0.9003 \cos\theta$	$\pi/4 - 3\pi/4$
	$- 1 + 0.9003 \cos\theta$	$3\pi/4 - \pi$

Table 10-8. Equations for Circular Ring Under Four Way
Cosine Load Distribution, Fig. 10-14

Symbol	Function	Range of Application
K_u	$- 0.1111 \cos 2\theta$	$0 - \pi$
K_ψ	$0.2222 \sin 2\theta$	$0 - \pi$
K_M	$0.3333 \cos 2\theta$	$0 - \pi$

Table 10-9. Equations for Circular Ring Under Cosine
Loading and Tensile Reaction, Fig. 10-15

Symbol	Function	Range of Application
K_u	$0.1989\ \theta \sin\theta + (0.4081 - 0.0796\ \theta^2) \cos\theta - 0.3618$	$0 - \pi/2$
	$(0.3750 - 0.0398\ \theta) \sin\theta + (0.3658 - 0.2500\ \theta) \cos\theta - 0.3618$	$\pi/2 - \pi$
K_ψ	$(0.0796\ \theta^2 - 0.2092) \sin\theta + 0.0398\ \theta \cos\theta$	$0 - \pi/2$
	$(0.2500\ \theta - 0.4055) \sin\theta + (0.1250 - 0.0398\ \theta) \cos\theta$	$\pi/2 - \pi$
K_M	$0.3183\ \theta \sin\theta + 0.2387 \cos\theta - 0.3618$	$0 - \pi/2$
	$0.5000 \sin\theta - 0.0796 \cos\theta - 0.3618$	$\pi/2 - \pi$

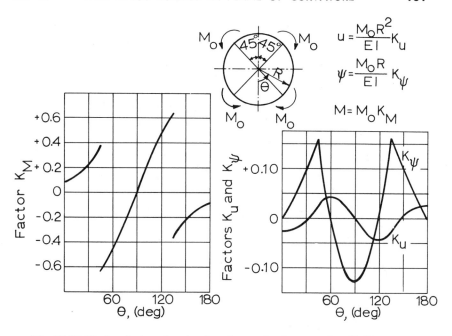

Fig. 10-13. Design factors for circular ring under four couples, Table 10-7

$$u = \frac{q_m R^4}{EI} K_u \qquad \psi = \frac{q_m R^3}{EI} K_\psi \qquad M = q_m R^2 K_M$$

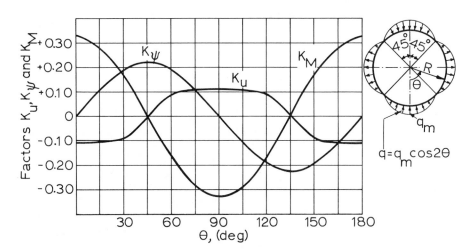

Fig. 10-14. Design factors for circular ring under four way cosine load distribution, Table 10-8

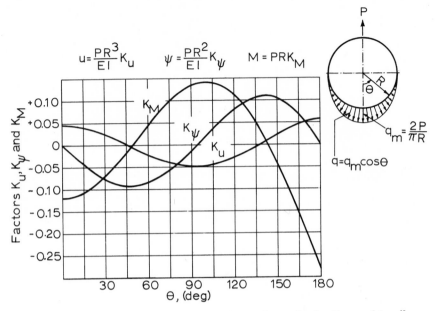

Fig. 10-15. Design factors for circular ring under cosine loading and tensile reaction, Table 10-9

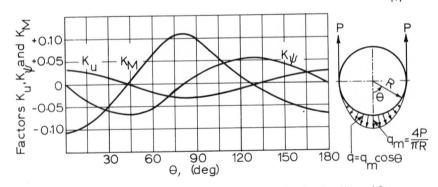

Fig. 10-16. Design factors for circular ring under cosine loading with two tangential reactions, Table 10-10

Table 10-10. Equations for Circular Ring Under Cosine Loading
with Two Tangential Reactions, Fig. 10-16

Symbol	Function	Range of Application
K_u	$0.5570\,\theta \sin\theta + (0.9382 - 0.1592\,\theta^2)\cos\theta$ -0.9053	$0 - \pi/2$
	$(0.0796\,\theta - 0.2500)\sin\theta + 0.0681\cos\theta$ $+0.0947$	$\pi/2 - \pi$
K_ψ	$(0.1592\,\theta^2 - 0.3812)\sin\theta + 0.2387\,\theta\cos\theta$	$0 - \pi/2$
	$0.0115\sin\theta + (0.0796\,\theta - 0.2500)\cos\theta$	$\pi/2 - \pi$
K_M	$0.6366\,\theta \sin\theta + 0.7958\cos\theta - 0.9053$	$0 - \pi/2$
	$0.1592\cos\theta + 0.0947$	$\pi/2 - \pi$

Table 10-11. Equations for Circular Ring Under Cosine Loading
with Two Radial Reactions, Fig. 10-17

Symbol	Function	Range of Application
K_u	$0.6309\cos\theta + 0.3697\,\theta\sin\theta - 0.1125\,\theta^2\cos\theta$ $- 0.6049$	$0 - \pi/2$
	$(0.5303 + 0.0321\,\theta)\sin\theta + (0.5709$ $- 0.3535\,\theta)\cos\theta - 0.6049$	$\pi/2 - 3\pi/4$
	$(1.0098 - 0.3214\,\theta)\sin\theta - 0.6157\cos\theta$ $- 0.6049$	$3\pi/4 - \pi$
K_ψ	$(0.1125\,\theta^2 - 0.2611)\sin\theta + 0.1447\,\theta\cos\theta$	$0 - \pi/2$
	$(0.3535\,\theta - 0.5388)\sin\theta + (0.0321\,\theta$ $+ 0.1768)\cos\theta$	$\pi/2 - 3\pi/4$
	$0.2942\sin\theta + (1.0098 - 0.3214\,\theta)\cos\theta$	$3\pi/4 - \pi$
K_M	$0.4501\,\theta\sin\theta + 0.5144\cos\theta - 0.6049$	$0 - \pi/2$
	$0.7071\sin\theta + 0.0642\cos\theta - 0.6049$	$\pi/2 - 3\pi/4$
	$- 0.6429\cos\theta - 0.6049$	$3\pi/4 - \pi$

Numerical Examples

Design Problem 18: Find the ultimate load capacity and deflection of a proving ring shown in Fig. 10-18, assuming yield strength $S_y = 100,000$ psi and the modulus of elasticity $E = 29.5 \times 10^6$ psi.

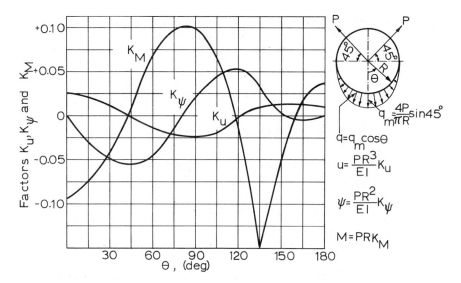

Fig. 10-17. Design factors for circular ring under cosine loading with two radial reactions, Table 10-11

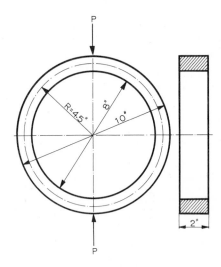

Fig. 10-18. Proving ring

Solution: The maximum bending moment may be found from Table 10-2, or design curve in Fig. 10-8, for $\theta = 0$. The same result is obtained also from Eq. (232), yielding $M = PR/\pi$. In pure bending, Eq. (8), the stress is $S_y = M/Z$. Since for the rectangular section

$Z = bh^2/6$, we get

$$S_y = \frac{6PR}{\pi bh^2}$$

from which

$$P = \frac{\pi bh^2 S_y}{6R}$$

Substituting $I = bh^3/12$ in deflection formula, Eq. (234), gives

$$Y = \frac{3PR^3 \, (\pi^2 - 8)}{\pi E bh^3}$$

Substituting for P, in the above formula, yields

$$Y = \frac{(\pi^2 - 8) \, S_y R^2}{2Eh}$$

Hence introducing the numerical data, gives

$$Y = \frac{(\pi^2 - 8) \times 100{,}000 \times 4.5^2}{2 \times 29.5 \times 10^6 \times 1^2}$$

$$Y = 0.0638 \text{ in.}$$

The corresponding ultimate load is

$$P = \frac{\pi \times 2 \times 1^2 \times 100{,}000}{6 \times 4.5}$$

$$P = 23{,}300 \text{ lb}$$

Design Problem 19: A machine frame is supported tangentially at two points as shown in Fig. 10-19 and carries a variable radial loading distributed over one half of the circumference according to cosine load. If the mean frame radius $R = 15$ inches and the frame has a rectangular cross-section measuring 4×0.5 inches, calculate radial deflection and the corresponding bending stress at the bottom of the frame, assuming that each tangential load $P = 2000$ lbs, and $E = 30 \times 10^6$ psi.

Solution: Moment of inertia $I = \dfrac{bh^3}{12} = \dfrac{4 \times 0.5^3}{12} = 0.0417 \text{ in}^4$

Radial deflection at the bottom of the ring is given in Fig. 10-16, as $u = \dfrac{PR^3}{EI} K_u$. For $\theta = 0$, Table 10-10 gives 0.0329. Hence

$$u = \frac{2000 \times 15^3}{30 \times 10^6 \times 0.0417} \; 0.0329 = 0.1777 \text{ inches}$$

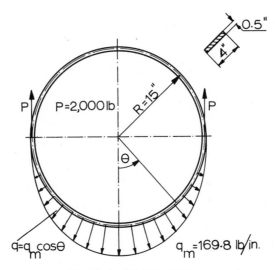

Fig. 10-19. Machine frame

The bending moment is found from Fig. 10-16 and Table 10-10 as follows

$$M = PR\,K_M$$

for $\theta = 0$, Table 10-10 yields $K_M = 0.7958 - 0.9053 = -0.1095$

Then

$$M = 2000 \times 15 \times (-0.1095)$$

$$M = -3285 \text{ lb-in.}$$

Therefore the bending stress is

$$S_b = -3285 \times 0.25/0.0417 = -19,700 \text{ psi} \quad \begin{array}{l}\text{(Outer ring} \\ \text{surface in} \\ \text{tension)}\end{array}$$

Design Problem 20: A support bracket made in form of a circular frame with rigid trunnions along the horizontal diameter, carries a vertical loading the distribution of which may be approximated by a cosine function as shown in Fig. 10-20. Calculate the net change in vertical ring diameter and the angle of slope of rigid trunnions due to frame deformation assuming that the maximum unit load at the bottom of the frame is 200 lb/in. Take the modulus of elasticity $E = 30 \times 10^6$ psi.

Solution: This case can be solved by the method of superposition, combining the solutions depicted in Figs. 10-12 and 10-16. According to Fig. 10-12, there is zero net change in vertical diameter under the

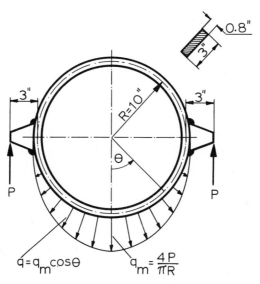

Fig. 10-20. Circular support bracket

action of two diametrically opposed bending couples. The radial deflection from Fig. 10-16 is

$$u = \frac{PR^3}{EI} K_u$$

The design curve for K_u, indicates that the values at $\theta = 0$ and at $\theta = 180$ deg are of the same sign. From Table 10-10,

$$K_u = 0.9382 - 0.9053 = 0.0329$$

$$\text{at } \theta = 0$$

and

$$K_u = -0.0681 + 0.0947 = 0.0266$$

$$\text{at } \theta = 180$$

Hence the total design factor

$$K_u = 0.0329 + 0.0266 = 0.0595$$

According to Fig. 10-16, $q_m = \dfrac{4P}{\pi R}$. Since $q_m = 200$ lb/in., the value of P, can be now calculated

$$P = \frac{\pi R q_m}{4} = \frac{\pi \times 10 \times 200}{4}$$

$$P = 1570 \text{ lbs}$$

The moment of inertia $I = 3 \times 0.8^3/12 = 0.128$ in.[4] Hence the required net change in vertical diameter is

$$u = \frac{1570 \times 10^3 \times 0.0595}{30 \times 10^6 \times 0.128}$$

$$u = 0.0243 \text{ in.}$$

The slope at the attachment point of the trunnion, $\theta = 90$ deg, is made up of two components. Due to offset of reactions P, Fig. 10-20, the ring becomes pear-shaped under load, developing a flatter shape in the lower part of the ring, thereby causing the tangent to the mean circle at $\theta = 90$ deg to rotate counterclockwise. The effect of cosine loading alone, supported as shown in Fig. 10-16, is to develop similar pear-shaped configuration with flatter portion of the ring at the top, thereby causing the tangent to rotate clockwise. Hence to get net angle of slope for the trunnions the value of K_ψ should be taken as the algebraic difference of the cases illustrated in Figs. 10-12 and 10-16.

Negative rotation (Table 10-6.) $K_\varphi = -0.1592$

$\theta = \pi/2$

Positive rotation (Table 10-10.) $K_\varphi = +0.0115$

$\theta = \pi/2$

Hence the net angle of rotation is given by the following:

$$\psi = 0.0115 \frac{PR^2}{EI} - 0.1592 \frac{M_o R}{EI}$$

In the case of Fig. 10-20, $M_o = 3 \times 1570 = 4710$ lb-in. Therefore substituting the relevant numerical data gives

$$\psi = \frac{0.0115 \times 1570 \times 10^2}{30 \times 10^6 \times 0.128} - \frac{0.1592 \times 4710 \times 10}{30 \times 10^6 \times 0.128}$$

$$\psi = -0.0015 \text{ rad (counterclockwise rotation)}$$

Symbols for Chapter 10

A	Area of cross-section, in.[2]
b	Width of rectangular section, in.
E	Modulus of elasticity, psi
G	Modulus of rigidity, psi

g	Acceleration due to gravity, in/sec^2
H	Horizontal load, lb
h	Depth of cross-section, in.
I	Moment of inertia, in.^{4}
K_M	Factor for bending moment
K_u	Factor for radial deflection
K_ψ	Factor for slope
M	Bending moment, lb-in.
M_f	Fixing moment, lb-in.
M_o	Externally applied bending couple, lb-in.
M_q	Bending moment due to uniform load, lb-in.
M_1, M_2	Bending moments for various portions, lb-in.
N	Normal force, lb
P	Vertical load, lb
Q	Transverse shearing force, lb
q	Uniform load, lb/in.
q_m	Maximum load per unit length, lb/in.
R	Mean radius of curvature, in.
S	Stress, psi
S_b	Bending stress, psi
U	Elastic strain energy, lb-in.
u	Radial displacement, in.
V	Peripheral velocity, in/sec
W	Radial load, lb.
w	Weight per unit length, lb/in.
X	Horizontal deflection, in.
Y	Vertical deflection, in.
Z	Section modulus, in.^{3}

β   Angle at which load is applied, rad

γ   Specific weight, $lb/in.^3$

ε   Auxiliary angle, rad

θ   Angle at which forces are considered, rad

ψ   Slope, rad

$χ = R/h$ Ratio of radius of curvature to depth of section

CHAPTER ELEVEN

Transversely Loaded Rings

Assumptions

The design of machine members and various supporting hardware frequently involves the calculation of stress and deformations of a circular elastic ring loaded normal to the plane of curvature. The axis of the ring is usually considered to be a plane curve and it is assumed that the ring maintains its circularity under load. Since the forces act normal to the plane of curvature the bending and twisting effects are analyzed simultaneously. An extensive theoretical treatment of out-of-plane deformation of circular rings is given by Biezeno and Grammel (Ref. 27). This chapter treats several cases of rings with statically determinate supports which have varied applications (Ref. 22). The analysis of these members is generally considered to be beyond the scope of most text books on strength of materials and engineering handbooks.

As before, the rings dealt with in this chapter are assumed to be thin and uniform in cross-section. The deformation is therefore caused mainly by bending and twisting moments, and the effects of shear and normal stresses are neglected.

Twist of a Circular Ring

When a free circular ring is subjected to the action of two diametrically opposed twisting moments the deflection formulas can be derived with the aid of Eqs. (42) and (43). An example of the derivation procedure, obtained from the published account Ref. 22, is given below.

Consider a free ring under toroidal moments, illustrated in Fig. 11-1. The plan view of one half of the ring indicates the geometrical

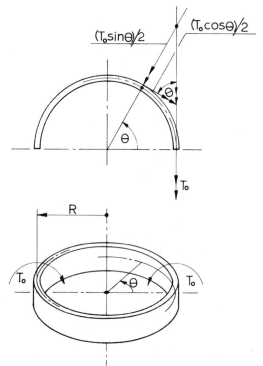

Fig. 11-1. Circular ring under toroidal moments

relations necessary for establishing the bending and twisting moments. It follows then that

$$M = (T_o \sin \theta)/2 \text{ *}$$ (282)

and

$$T = (T_o \cos \theta)/2$$ (283)

From Eq. (282),

$$\frac{dM}{d\theta} = \frac{T_o \cos \theta}{2}$$ (284)

Hence substituting Eqs. (282), (283) and (284) into Eq. (42), gives

$$\frac{d^3Y}{d\theta^3} + \frac{dY}{d\theta} = A \cos \theta \qquad (285)$$

Where

$$A = -\frac{R^2 T_o}{2EI} (1 + \lambda) \qquad (286)$$

Here, as before, λ, represents the ratio of flexural to torsional rigidity EI/GK. This parameter is important in all studies of out-of-plane deformation of curved members. The auxiliary relation for the solution of the differential equation, Eq. (285), is

$$n (n^2 + 1) = 0 \qquad (287)$$

It is evident that the roots of this equation are 0, $+i$, and $-i$. Therefore following the usual procedure in the solution of the general linear equation, given in standard books on calculus, the complementary function is

$$F_c = A_1 + A_2 \cos \theta + A_3 \sin \theta \qquad (288)$$

In Eq. (288), A_1, A_2 and A_3 denote arbitrary integration constants to be determined from the geometric boundary conditions. Since the complete solution of the linear differential equation requires that the complementary function F_c, be added to the particular integral, we find by observation that

$$F_P = -A (\sin \theta + \theta \cos \theta)/2 \qquad (289)$$

Although several rigorous mathematical methods are available for finding particular integrals, with some experience the observation method can give relatively quick results. In this approach we assume the most likely form of the particular integral and verify the result by differentiation. For instance in the problem under consideration, making $Y = F_P$, and obtaining the appropriate derivatives from Eq. (289), gives

$$\frac{dY}{d\theta} = -\frac{A}{2} (2 \cos \theta - \theta \sin \theta)$$

and

$$\frac{d^3Y}{d\theta^3} = -\frac{A}{2} (\theta \sin \theta - 4 \cos \theta)$$

---

* For meaning of symbols and dimensional units involved for this and other equations in this chapter see material at end of chapter.

Simple addition of these two equations in accordance with Eq. (285) proves that the selected particular integral, Eq. (289) fulfills the necessary requirement.

Hence the general solution of the differential equation, Eq. (285), is

$$Y = F_c + F_P \tag{290}$$

Here F_c and F_P are given by Eqs. (288) and (289), respectively. Combining these equations gives

$$Y = A_1 + A_2 \cos \theta + \left(A_3 - \frac{A}{2} \right) \sin \theta - \frac{A\theta \cos \theta}{2} \tag{291}$$

Looking over the diagram in Fig. 11-1, we observe that when $\theta = 0$, the out-of-plane deflection at that point must be zero. Hence putting $\theta = 0$, and $Y = 0$ in Eq. (291), gives $A_1 = -A_2$. Because of symmetry of deformation of the ring the bending slope at $\theta = 0$ and $\theta = \pi/2$ must be equal to zero. To find the slope equation, differentiate Eq. (291) with respect to θ and make $A_1 = -A_2$. This yields

$$\frac{dY}{d\theta} = (A_3 - A) \cos \theta + \frac{A\theta \sin \theta}{2} - A_2 \sin \theta \tag{292}$$

Hence, putting $\theta = 0$ and $\theta = \pi/2$, in Eq. (292), and making $dY/d\theta = 0$, gives two additional equations from which $A_2 = \pi A/4$ and $A_3 = A$. Substituting for A_2, A_3 and A, as well as $A_1 = -A_2$ in Eq. (291) gives the design formula for the deflection at any point of the ring shown in Fig. 11-1.

$$Y = \frac{T_o R^2}{8EI} (1 + \lambda) (2\theta \cos \theta - \pi \cos \theta + \pi - 2 \sin \theta) \tag{293}$$

It is noted that for $\theta = 0$, Eq. (293), vanishes in accordance with the prescribed boundary conditions. The maximum value of the deflection is obtained from Eq. (293), when $\theta = \pi/2$ is substituted.

$$Y = \frac{T_o R^2}{8EI} (\pi - 2) (1 + \lambda) \tag{294}$$

The slope due to the transverse bending can be obtained by differentiating deflection, Eq. (293) with respect to θ and substituting the result in Eq. (54). This yields

$$\psi = \frac{T_o R}{8EI} (1 + \lambda) (\pi - 2\theta) \sin \theta \tag{295}$$

Again, in accordance with the prescribed boundary conditions of the problem the slope is found to be zero when $\theta = 0$, or $\theta = \pi/2$ is substituted in Eq. (295).
The design formula for the angle of twist can be now found from Eq. (43). Hence differentiating Eq. (293), with respect to θ and substituting the result together with Eq. (282) into Eq. (43), gives

$$\eta = \frac{T_o R}{8EI} [2 (1 - \lambda) \sin \theta + (1 + \lambda) (\pi - 2\theta) \cos \theta] \quad (296)$$

Substituting in turn $\theta = 0$ and $\theta = \pi/2$ in Eq. (296), yields

$$\eta = \pi T_o R (1 + \lambda)/8EI \quad (297)$$

and

$$\eta = T_o R (1 - \lambda)/4EI \quad (298)$$

Since by the definition $\lambda = EI/GK$ is always greater than one, Eqs. (297) and (298) disclose an interesting phenomenon. When a ring is twisted under the action of toroidal moments, Fig. 11-1, midway between the moments applied the ring section actually rotates in the opposite sense to T_o. This observation can be easily verified by testing a simple paper model.

Ring Under Transverse Uniform Load

Circular ring subjected to transverse uniform load is shown in Fig. 11-2. The vertical reactions here are statically determinate and are equal to $\pi q R$, each. It is assumed that the ring cross-sections can rotate freely at the supports. The bending and twisting moments, existing at an arbitrary cross-section defined by θ, can be obtained

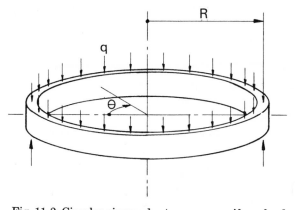

Fig. 11-2. Circular ring under transverse uniform load

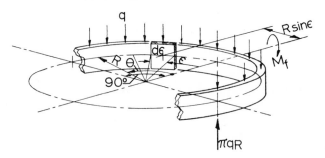

Fig. 11-3. Auxiliary diagram for transversely loaded circular ring

with the aid of the diagram in Fig. 11-3. Consider for instance a hypothetical cut of the ring at $\theta = 0$. To keep this end of the member in equilibrium the vertical reaction for one quarter of the ring is $\pi qR/2$. The fixing moment M_f supporting one quarter of the ring makes the bending slope at $\theta = 0$, equal to zero. Its value follows from the summation of all elementary moments about the horizontal axis shown in the diagram. In mathematical terms this is

$$M_f = \int_0^{\pi/2} (R \sin \varepsilon)\, qR d\varepsilon = qR^2 \qquad (299)$$

The bending moment caused by the uniformly distributed load M_q, about an arbitrary section defined by θ, is given by Eq. (135) as before. Hence the total bending moment at θ becomes

$$M = M_q - M_f \cos \theta + \frac{\pi qR^2 \sin \theta}{2} \qquad (300)$$

Substituting Eqs. (135) and (299) into Eq. (300), in accordance with the adopted sign convention, gives

$$M = qR^2 \left(\frac{\pi}{2} \sin \theta - 1 \right) \qquad (301)$$

It may be recalled that negative bending moment was assumed to produce tension in the upper surface of the ring. To verify the sign for Eq. (301), put $\theta = 0$, to get $M = -qR^2$.

Since the ring is assumed to roll freely at the supports there will be no fixing twisting moment at $\theta = 0$. The twisting moment due to the uniformly distributed load q is given by Eq. (136). The total twisting moment at an arbitrary cross-section defined by θ can be written as follows

$$T = T_q + M_f \sin \theta - \frac{\pi qR^2}{2}(1 - \cos \theta) \qquad (302)$$

Hence substituting Eqs. (136) and (299) in Eq. (302), yields

$$T = qR^2 \left[\theta + \frac{\pi}{2} (\cos \theta - 1) \right] \qquad (303)$$

It may be of interest to note that putting $\theta = 0$, $\theta = \pi/2$ or $\theta = \pi$, in Eq. (303), indicates no twist of the ring cross-section. The maximum value of the twisting moment can be found by the usual rule of differential calculus dealing with maximizing a given function. Therefore, obtaining $dT/d\theta$ from Eq. (303) and making the result equal to zero, gives

$$\frac{dT}{d\theta} = qR^2 \left(1 - \frac{\pi}{2} \sin \theta \right) = 0$$

From this, $\sin \theta = 2/\pi$, and θ is found to be approximately $39\frac{1}{2}$ degrees.

Once the general expressions for the bending and twisting moments, Eqs. (301) and (303) are established, the procedure utilized in the development of the design formulas for the case of toroidal moments can be applied to find the deflection, slope and the angle of twist at any point of the ring shown in Fig. 11-2. This yields

$$Y = \frac{qR^4}{8EI} \{ 2\pi \, (\theta \cos \theta - \sin \theta) + \pi^2 \, (1 - \cos \theta) + \lambda \, [2\pi \, (\theta \cos \theta$$

$$- 3 \sin \theta) + \pi^2 \, (1 - \cos \theta) + 4\theta \, (\pi - \theta)] \} \qquad (304)$$

$$\psi = \frac{qR^3}{8EI} \{ \pi \, (\pi - 2\theta) \sin \theta + \lambda \, [\, \pi \, (\pi - 2\theta) \sin \theta$$

$$+ 4 \, (\pi - 2\theta - \pi \cos \theta)] \} \qquad (305)$$

and

$$\eta = \frac{\pi qR^3}{8EI} (1 + \lambda) \, [2 \sin \theta + (\pi - 2\theta) \cos \theta - 8/\pi] \qquad (306)$$

**Transverse Deflection Under Uniform Load
by Method of Castigliano**

The preceding solution by the method of differential equations is especially useful when the general displacement formulas are required. The same results can be obtained by the method of fictitious loading and the theorem of Castigliano. To illustrate this point, calculate the maximum deflection for the case of the transverse uniform load shown in Fig. 11-2. Consider the equilibrium of one half of the ring subjected to uniform load q and fictitious load P as illustrated in Fig. 11-4. The

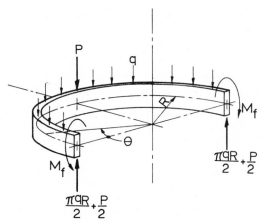

Fig. 11-4. Half-ring equilibrium under transverse uniform loading

relevant bending and twisting moments for this case are

$$M = qR^2 \left(\frac{\pi \sin \theta}{2} - 1 \right) + \frac{PR}{2} (\sin \theta - \cos \theta) \qquad (307)$$

and

$$T = qR^2 \left(\theta + \frac{\pi}{2} \cos \theta - \frac{\pi}{2} \right) + \frac{PR}{2} (\sin \theta + \cos \theta - 1) \qquad (308)$$

The vertical deflection at, $\theta = \pi/2$, follows from Eq. (53)

$$Y = \frac{2R}{EI} \int_0^{\pi/2} M \frac{\partial M}{\partial P} \, d\theta + \frac{2R}{GK} \int_0^{\pi/2} T \frac{\partial T}{\partial P} \, d\theta \qquad (309)$$

From Eqs. (307) and (308)

$$\frac{\partial M}{\partial P} = \frac{R}{2} (\sin \theta - \cos \theta) \qquad (310)$$

and

$$\frac{\partial T}{\partial P} = \frac{R}{2} (\sin \theta + \cos \theta - 1) \qquad (311)$$

Hence substituting Eqs. (307), (308), (310) and (311) into Eq. (309) and integrating yields

$$Y = \frac{\pi q R^4}{8EI} [\pi - 2 + 2\lambda (\pi - 3)] \qquad (312)$$

Note that as before, P is made equal to zero in bending and twisting moment equations, but only after the required partial derivatives $\partial M/\partial P$ and $\partial T/\partial P$ had been found. When $\theta = \pi/2$ is introduced into Eq. (304), Eq. (312) is readily obtained. On the other hand when $\theta = 0$ is substituted in Eq. (304) the deflection vanishes in line with the boundary conditions pertinent to the case shown in Fig. 11-2.

Ring on Multiple Supports

A useful summary of design equations for a circular ring under a combined transverse loading and multiple supports is taken from the paper on rings and arcuate beams (Ref. 22). The original equations, forming the basis for this section, have been developed by McGuiness, (Ref. 28).

The general view of the ring subjected to transverse uniform and concentrated loading is given in Fig. 11-5. The corresponding working formulas are

Vertical reaction at any support

$$V = (P + 2\pi qR)/n \tag{313}$$

Bending moment at $\theta = 0$

$$M = -(PRD_1 + qR^2D_2)/12n^2 \tag{314}$$

Bending moment at $\theta = \pi/n$

$$M = (PRD_1 + qR^2D_3)/12n^2 \tag{315}$$

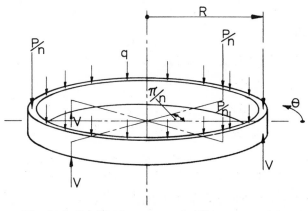

Fig. 11-5. Combined transverse loading on n supports

Twisting moment at $\theta = \pi/2n$

$$T = (P + \pi qR)\, RD_4/2 \tag{316}$$

and transverse deflection at $\theta = \pi/n$

$$Y = \frac{R^3}{n^4}(P + \pi qR)\left(\frac{D_5}{EI} + \frac{D_6}{GK}\right) \tag{317}$$

The numerical design factors for the solution of the above equations are given in Table 11-1, to help the designer in his work. Tabakman and Valentijn (Ref. 35) solved a similar problem using differential equations, Eqs. (39), (40) and (42) and presented design formulas for bending moments, twisting moments, deflection, slope and the angle of twist.

It should be noted here however, that these results do not apply to the case of a partial ring with the ends fixed as to slope and rotation.

Table 11-1. Force and Displacement Factors for Combined Loading

n	D_1	D_2	D_3	D_4	D_5	D_6
2	12.000	47.9909	27.2984	0.20710	2.2832	0.56640
3	10.3920	42.7030	22.5968	0.05157	1.6305	0.17950
4	9.9408	41.2038	21.2623	0.02060	1.4676	0.09072
5	9.7476	40.5567	20.6968	0.01031	1.4006	0.05518
6	9.6462	40.2174	20.3914	0.005883	1.3659	0.03740
7	9.5847	40.0100	20.2142	0.003674	1.3455	0.02859
8	9.5472	39.8814	20.0861	0.002450	1.3327	0.01999
9	9.5202	39.7989	20.0182	0.001711	1.3239	0.01587
10	9.5031	39.7386	19.9692	0.001250	1.3178	0.01296
11	9.4896	39.6821	19.9422	0.0009345	1.3135	0.01075
12	9.4824	39.6569	19.9051	0.0007166	1.3097	0.00855
36	9.4323	39.4835	19.8008	0.00002646	1.2952	0.00154
limit	3π	$4\pi^2$	$2\pi^2$	0	$\pi^3/24$	0

The maximum transverse deflection for the case shown in Fig. 11-2, can be now checked with the aid of McGuiness formula, Eq. (317).

Substituting $n = 2$, $P = 0$, $EI/GK = \lambda$ and the relevant displacement factors D_5 and D_6 from Table 11-1, gives previously derived deflection formula, Eq. (312).

Four Point Loading

To aid in the analysis of gyroscope gimbals, handling equipment and other similar systems, working equations for the circular ring, loaded and supported as shown in Fig. 11-6, reprinted from Ref. 22, are given below.

$$M = \frac{PR}{2} (\sin \theta - \cos \theta) \tag{318}$$

$$T = \frac{PR}{2} (\sin \theta + \cos \theta - 1) \tag{319}$$

$$Y = \frac{PR^3}{8EI} \left\{ 2 (\theta - 1) \sin \theta + (2\theta + 2 - \pi) \cos \theta + \pi - 2 \right.$$
$$\left. + \lambda [(6 + 2\theta - \pi) \cos \theta - 2 (3 - \theta) \sin \theta + \pi + 4\theta - 6] \right\} \tag{320}$$

$$\psi = \frac{PR^2}{8EI} \left\{ 2 \theta \cos \theta + (\pi - 2\theta) \sin \theta \right.$$
$$\left. + \lambda [(\pi - 2\theta - 4) \sin \theta + 2 (\theta - 2) \cos \theta + 4] \right\} \tag{321}$$

$$\eta = \frac{PR^2}{8EI} (1 + \lambda) [2 (1 - \theta) \sin \theta + (\pi - 2 - 2\theta) \cos \theta] \tag{322}$$

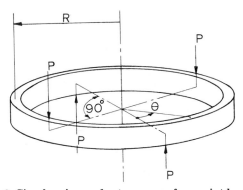

Fig. 11-6. Circular ring under transverse four point loading

When $\theta = \pi/2$, Eq. (320) gives essentially the same result as that found with the aid of Eq. (317) when $q = 0$ and P/n, from Fig. 11-5, is taken to be equal to P shown in Fig. 11-6.

Ring on Trunnion Supports

The method of superposition can be conveniently applied to important design problems involving transversely loaded rings supported eccentrically at trunnions.

Consider, for instance a trunnion-supported continuous ring, carrying a uniformly distributed load, as shown in Fig. 11-7. The required design formulas can be obtained by super-imposing the response of a ring under uniform load, Fig. 11-2, and that shown in Fig. 11-1 under toroidal moments. The toroidal moment T_0 follows from Fig. 11-7

$$T_o = Va = \pi Rqa$$

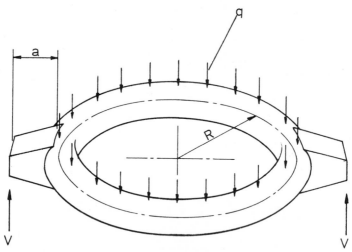

Fig. 11-7. Trunnion-supported continuous ring

Hence combining Eqs. (293) and (304), and substituting for T_o, gives

$$Y = \frac{qR^3}{8EI}\left[\pi\,(a+R)\,(1+\lambda)\,(2\theta\cos\theta - 2\sin\theta - \pi\cos\theta + \pi)\right.$$

$$\left. + 4\lambda R\,(\pi\theta - \theta^2 - \pi\sin\theta)\right] \tag{323}$$

Similarly the slope is obtained from Eqs. (295) and (305).

$$\psi = \frac{qR^2}{8EI}\left[\pi\,(a+R)\,(1+\lambda)\,(\pi-2\theta)\,\sin\theta + 4\lambda R\,(\pi-2\theta-\pi\cos\theta)\right]$$

$$\text{(324)}$$

Finally the angle of twist following from superimposing Eq. (296) and (306), yields

$$\eta = \frac{qR^2}{8EI}\Bigg\{ 2\pi\sin\theta\,[R\,(1+\lambda)+a\,(1-\lambda)]$$
$$+ \pi\,(\pi-2\theta)\,(a+R)\,(1+\lambda)\cos\theta - 8R\,(1+\lambda)\Bigg\}$$

$$\text{(325)}$$

The corresponding bending and twisting moments are

$$M = \frac{\pi qR\sin\theta}{2}\,(a+R) - qR^2 \qquad \text{(326)}$$

and

$$T = \frac{\pi qR\cos\theta}{2}\,(a+R) + qR^2\left(\theta-\frac{\pi}{2}\right) \qquad \text{(327)}$$

The maximum transverse deflection follows from Eq. (323), when $\theta = \pi/2$ is substituted. This gives

$$Y = \frac{\pi qR^3}{8EI}\left[(a+R)\,(1+\lambda)\,(\pi-2) + R\lambda\,(\pi-4)\right] \qquad \text{(328)}$$

When eccentricity a becomes zero, Eq. (328) reduces to Eq. (312). With the aid of the basic cases presented in this chapter, numerous design situations can be analyzed by applying the principle of superposition. To minimize the displacements, trunnion support points may be moved inside the ring. In such a case the sign of a in the above formulas should change to negative.

Numerical Examples

Design Problem 21: A continuous machine frame, formed in a circle with mean radius of $R = 15$ inches, is designed to be supported on two diametrically opposed trunnions. It carries a uniformly distributed load $q = 50$ lbs/inch of circumference, acting normal to the plane of curvature as shown in Fig. 11-7. Assuming the frame cross-section to be tubular with outer and inner diameters of 3 and 2.75 inches respectively, calculate the maximum deflections for the offset lugs being placed outside or inside the frame ring. Assume the offset lugs to be perfectly rigid and $a = 3$ in. Take $E = 30 \times 10^6$ psi and $G = 0.4E$.

Solution: Since the tubular frame in this case is relatively thin calculate the moment of inertia from the approximate formula. This gives

$$I = \pi r^3 t = 3.14 \times 2.875^3 \times 0.125^2$$

$$I = 1.17 \text{ in.}^4$$

Since for a hollow circular cross-section the torsional shape factor is actually twice the value of I, then

$$\lambda = EI/(0.4E \times 2I) = 1.25$$

Equation (328) can now be used to evaluate the maximum deflections. When a is positive, offset lugs are placed outside and Eq. (328) gives

$$Y = \frac{3.14 \times 50 \times 15^3}{8 \times 30 \times 10^6 \times 1.17} \Big[(3 + 15) \, (1 + 1.25) \, (3.14 - 2) $$
$$ + 1.25 \times 15 \, (3.14 - 4) \Big]$$

$$Y = 0.0565 \text{ in.}$$

When offset lugs are placed inside the ring, Eq. (328) applies provided the term $(a + R)$ is replaced with the term $(R - a)$. This yields

$$Y = 0.0277 \text{ in.}$$

Design Problem 22: Develop a theoretical relationship between the offset to radius ratio a/R and the ratio of flexural to torsional rigidity λ, for which the maximum vertical displacement Y, Fig. 11-7, can be made equal to zero, when the lug is on the inside of the ring.

Solution: Consider modified version of Eq. (328), in which $(a + R)$ is replaced by $(R - a)$. For Y to be zero the expression in the brackets must be equal to zero.

$$(R - a) \, (1 + \lambda) \, (\pi - 2) + R\lambda \, (\pi - 4) = 0$$

Dividing this expression by R and solving for a/R, gives

$$\frac{a}{R} = \frac{\pi - 2 + 2\lambda \, (\pi - 3)}{(1 + \lambda) \, (\pi - 2)}$$

As stated before, λ must always be greater than 1. Taking the lowest practical value $\lambda = 1.25$, the above formula gives

$$\frac{a}{R} = \frac{1.14 + 2.5 \, (3.14 - 3)}{2.25 \times 1.14} = 0.582$$

For a relatively high ratio λ, the offset to radius ratio may be approximated by

$$\frac{a}{R} = \frac{1}{\lambda} + \frac{2\,(\pi - 3)}{\pi - 2}$$

Hence the theoretical minimum value of a/R is about 0.25. It should be noted here that such a low ratio of a/R may not be acceptable because of the tendency of thin cross-sectional members to become elastically unstable. However, this example indicates that at least partial reduction of the maximum deflection can be considered attainable in certain practical cases.

Design Problem 23: A closed circular ring of constant cross-section is resting on equidistant multiple supports and carries a transverse load uniformly distributed along the periphery. Evaluate the relative effect of the torsional rigidity on the maximum out-of-plane deflection. Assume $\lambda = 1.25$.

Solution: When $P = 0$, and $\lambda = 1.25$ is substituted in Eq. (317), we get the following simplified expression for out-of-plane deflection.

$$Y = \frac{\pi q R^4}{n^4 EI}\,(D_5 + 1.25\,D_6)$$

The contribution of torsional rigidity in the above equation is given by the term $1.25\,D_6$. Hence, utilizing Table 11-1, the following parameter can be evaluated as a function of the number of supports.

$$\frac{Y n^4 EI}{\pi q R^4} = D_5 + 1.25\,D_6$$

The results of this evaluation are plotted in Fig. 11-8. It is easy to see from this diagram that the effect of torsion on the maximum deflection of the ring on multiple supports becomes negligible when the number of supports is greater than about four. Also, Eq. (317) indicates that this effect is the same regardless whether the transverse loads are concentrated, uniformly distributed or a combination of both.

Symbols for Chapter 11

A	Auxiliary factor, in.
A_1, A_2, A_3	Integration constants
a	Offset, in.
$D_1, D_2, \ldots D_6$	Design factors

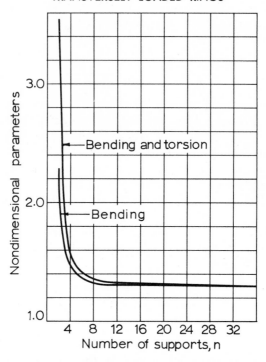

Fig. 11-8. Contribution of bending and torsion to deflection of ring on n supports

E	Modulus of elasticity, psi
F_c	Complementary function, in.
F_P	Particular solution, in.
G	Modulus of rigidity, psi
I	Moment of inertia, in.^{4}
K	Torsional shape factor, in.^{4}
M	Bending moment, lb-in.
M_f	Fixing moment, lb-in.
M_q	Bending moment due to uniform load, lb-in.
n	Number of equidistant supports
P	Vertical load, lb
q	Uniform load, lb/in.

R Mean radius of curvature, in.

r Mean radius of tube, in.

T Twisting moment, lb-in.

T_o Externally applied twisting couple, lb-in.

T_q Twisting moment due to uniform load, lb-in.

t Wall thickness, in.

V Vertical reaction, lb

Y Vertical deflection, in.

ε Auxiliary angle, rad

η Angle of twist, rad

θ Angle at which forces are considered, rad

λ Ratio of flexural to torsional rigidity

ψ Slope, rad

CHAPTER TWELVE

Theory on Neutral Axis

Assumptions

By the usually accepted definition neutral axis is the line created by the intersection of the neutral surface and the given cross-section along which the longitudinal fiber stress due to flexure is zero. When a curved member, having a small depth compared to the radius of initial curvature, is subject to pure bending the neutral axis can be assumed to pass through the centroid of the cross-section.

When a curved member of sharp curvature such as a machine hook, chain link or thick bearing ring is submitted to pure bending, plane sections are assumed to remain plane but the neutral axis is found to be displaced towards the center of curvature. The magnitude of such displacement depends on the shape and dimensions of the cross-section in relation to the mean radius of curvature.

The general mathematical relations, governing the method of calculation of the distance between the neutral and central axes, can be demonstrated for a curved beam of a rectangular cross-section subject to flexure (Ref. 29).

Shift of a Neutral Axis in a Rectangular Cross-section

According to Fig. 12-1, the neutral axis, defined by R_n, is assumed to be displaced towards the center of curvature by the amount, δ, while an arbitrary section corresponding to θ has rotated through an angle ψ. By the definition, the fiber at radius R_n retains its original

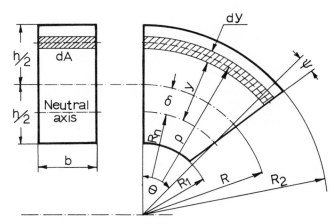

Fig. 12-1. Notation for finding displacement of neutral axis

length but any layer outside this radius must undergo elongation or shortening given by the amount $y\psi$. Since the originally unstrained length is $\varrho\theta$, the ratio of elongation to the unstrained length defines the strain

$$e = \frac{y\psi}{\varrho\theta} \quad *$$

Hence Hooke's law gives

$$S = eE = \frac{y\psi E}{\varrho\theta} \tag{329}$$

Equation (329) defines the normal stress on the cross-section resulting from bending.

The elementary normal force acting on the section dA, Fig. 12-1, is simply the product of the stress times the area. Utilizing Eq. (329), gives

$$SdA = \frac{y\psi E}{\varrho\theta} \, dA$$

Assuming that the modulus of elasticity E, does not vary across the section, the total normal force can be integrated as follows

$$N = \frac{\psi E}{\theta} \int_{R_1}^{R_2} \frac{y \, dA}{\varrho} \tag{330}$$

---

* For meaning of symbols and dimensional units involved for this and other equations in this chapter see material at end of chapter.

It is evident from Fig. 12-1, that the elementary area is

$$dA = bdy$$

For a given sectional geometry of the curved beam, R_n is constant, and y may be expressed in terms of radii

$$y = \varrho - R_n$$

Differentiating both sides of the above relation with respect to ϱ gives

$$dy = d\varrho$$

Introducing the above relations in Eq. (330) gives

$$N = \frac{\psi E}{\theta} \int_{R_1}^{R_2} \frac{\varrho - R_n}{\varrho} bd\varrho \qquad (331)$$

If the curved member is subjected to bending only, according to the conditions of equilibrium the net force N, normal to the cross-section must be zero. Hence Eq. (331) becomes

$$\frac{bE\psi}{\theta} \int_{R_1}^{R_2} d\varrho - \frac{bE\psi R_n}{\theta} \int_{R_1}^{R_2} \frac{d\varrho}{\varrho} = 0 \qquad (332)$$

Integrating Eq. (332) gives

$$R_2 - R_1 - R_n \log_e (R_2/R_1) = 0 \qquad (333)$$

Equation (333) is now solved for R_n and the following substitutions are made with reference to Fig. 12-1:

$$h = R_2 - R_1$$

$$\delta = R - R_n$$

$$R_2 = R + \frac{h}{2}$$

$$R_1 = R - \frac{h}{2}$$

This procedure yields the expression for the displacement of neutral axis towards the center of curvature for a rectangular cross-section

$$\delta = R - \frac{h}{\log_e \dfrac{(2R + h)}{(2R - h)}} \qquad (334)$$

When the value of h is rather small compared with the mean radius of curvature R, the second term of Eq. (334) may be expanded according to the well known theorem of Maclaurin. To illustrate the method of application of this useful theorem consider the logterm of Eq. (334).

$$f = \log_e \frac{2R + h}{2R - h} \tag{335}$$

The Maclaurin's series is expressed as follows

$$f(x) = f(0) + xf'(0) + \frac{x^2}{2!} f''(0) + \frac{x^3}{3!} f'''(0) + \dots \tag{336}$$

In Eq. (336) x denotes an independent variable and superscript primes denote the number of consecutive derivatives of the function with respect to x. In our case, h will be considered as an independent variable for the purpose of expanding the log-term given by Eq. (335). Taking only the first three derivatives of Eq. (335) gives

$$f'(h) = \frac{df}{dh} = \frac{4R}{4R^2 - h^2}$$

$$f''(h) = \frac{d^2f}{dh^2} = \frac{8Rh}{(4R^2 - h^2)^2}$$

$$f'''(h) = \frac{d^3f}{dh^3} = \frac{32R^3 + 24Rh^2}{(4R^2 - h^2)^3}$$

When $h = 0$, the above derivatives give

$$f'(0) = 1/R$$
$$f''(0) = 0$$
$$f'''(0) = 1/2R^3$$

Since $f(0) = 0$ and $f''(0) = 0$, the first and third terms of the Maclaurin's series, Eq. (336), drop out, while the second and fourth terms give

$$f = \frac{h}{R} + \frac{h^3}{12R^3} \tag{337}$$

Substituting this expansion in Eq. (334) yields

$$\delta = \frac{h^2R}{12R^2 + h^2} \tag{338}$$

In a similar manner other design formulas for the displacement of neutral axis in various cross-sections can be developed. Where the geometry of the cross-section cannot be simply expressed by mathematical relations graphical integration is usually employed.

Neutral Axis Correction for Bending Strain Energy

In calculating the elastic energy due to bending for a curved member of sharp curvature the usual expression $M^2/2EI$ can be modified to take into account the shift of the neutral axis. Taking moments about the center of curvature of the curved beam shown in Fig. 12-1, the elementary bending moment is

$$dM = S\varrho\, dA$$

Eliminating stress from the above relation with the aid of Eq. (329) gives

$$dM = \frac{y\psi E}{\theta}\, dA$$

The total moment of the normal forces is

$$M = \frac{\psi E}{\theta} \int_{R_1}^{R_2} y\, dA \qquad (339)$$

Equation (339) can be now rearranged through the following substitution

$$dA = b\, dy$$
$$y = \varrho - R_n$$
$$dy = d\varrho$$
$$dA = b\, d\varrho$$

Hence Eq. (339) becomes

$$M = \frac{bE\psi}{\theta} \int_{R_1}^{R_2} (\varrho - R_n)\, d\varrho$$

Integrating the above expression gives

$$M = \frac{AE\delta\psi}{\theta} \qquad (340)$$

The elementary work done by a couple M in rotating the beam cross-section through an angle $d\psi$, Fig. 12-2, is equal to the elastic energy of deformation provided this energy can be totally recovered upon the removal of external loading. This gives

$$dU = \frac{M\, d\psi}{2} \qquad (341)$$

From Eq. (340)

$$\psi = \frac{M\theta}{AE\delta}$$

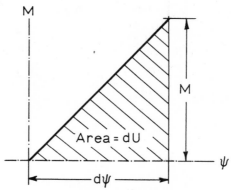

Fig. 12-2. Bending energy diagram

The rate of change of ψ, with respect to θ, follows from the above equation

$$d\psi = \frac{M \, d\theta}{AE\delta}$$

Substituting for $d\psi$, in Eq. (341), gives

$$dU = \frac{M^2}{2AE\delta} \, d\theta$$

Hence for the total curved member the elastic strain energy in bending becomes

$$U = \int \frac{M^2}{2AE\delta} \, d\theta \qquad (342)$$

The term defining the elastic strain energy, Eq. (342), enters the general energy and displacement expressions, Eqs. (46), (49) and (50).

Neutral Axis Correction for Bending Stresses

As pointed out in Chapter 2, the usual engineering formula $S_b = Mc/I$, representing a special case of a more general Winkler-Bach formula gives satisfactory results for straight beams. When the straight beam formula is applied to the curved members of sharp curvature such as hooks, chain links, thick rings, press frames and similar machine parts, the maximum bending stresses calculated on the assumption of straight beam theory may be too low. This is due to the shift of the neutral axis. Based on the theory of neutral axis and experimental analysis, Wilson and Quereau developed a simple correction factor to be used with the Winkler-Bach formula when calculating stresses in curved members, (Refs. 15 & 30). The effect of considerable initial curvature, reflected in the magnitude of the shift of the neutral axis, is such that the distribution of the normal stress over the depth of the cross-section is hyperbolic instead of linear, Fig. 12-3. The correction factor of Wilson and Quereau, de-

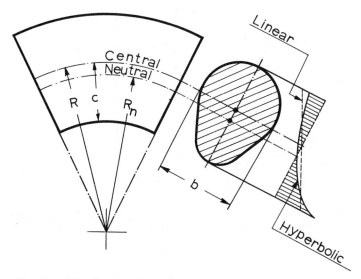

Fig. 12-3. Distribution of normal stresses in a thick curved member

noted in this book by W_q, may also be used when combined axial and bending loads are applied; ie.,

$$S = W_q \left(\frac{P}{A} + \frac{Mc}{I} \right) \tag{343}$$

When a curved bar has sharp curvature, Fig. 12-3, the critical stresses are found at the inside fibers. The approximate formula for W_q, due to Wilson and Quereau, applicable to inside fibers is

$$W_q = 1.00 + \frac{0.5I}{bc^2} \left(\frac{1}{R-c} + \frac{1}{R} \right) \tag{344}$$

Here I denotes the moment of inertia of bar cross-section about the centroidal axis, while b, c and R, are defined in Fig. 12-3.

Numerical Examples

Design Problem 24: For a curved member of deep cross-section, dimensioned in Fig. 12-4, calculate the ratios of the elastic strain energy due to bending using the formulas for thin and thick curved members, respectively. Find the shift of the neutral axis, using simplified and exact formulas.

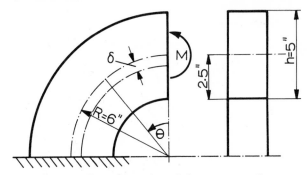

Fig. 12-4. Curved member of deep cross-section

Solution: Denote by U_1 *and* U_2 the elastic strain energy due to bending for thick and thin members, respectively. This gives

$$U_1 = \frac{1}{2AE\delta} \int_0^{\pi/2} M^2 \, d\theta$$

and

$$U_2 = \frac{R}{2EI} \int_0^{\pi/2} M^2 \, d\theta$$

Dividing both sides of the above equations yields

$$U_1/U_2 = I/A\delta R$$

Substituting

$$I = bh^3/12 \text{ and } A = bh$$

the above ratio becomes

$$U_1/U_2 = h^2/12R\delta$$

Substituting $h = 5''$ and $R = 6''$ in Eq. (334) gives

$$\delta = 6 - \frac{5}{\log_e \dfrac{(2 \times 6 + 5)}{(2 \times 6 - 5)}} = 0.438 \text{ in.}$$

Using simplified formula for the shift of the neutral axis, Eq. (338) yields

$$\delta = \frac{25 \times 6}{12 \times 36 + 25}$$

$$\delta = 0.328 \text{ in.}$$

Hence using exact formula for δ gives

$$U_1/U_2 = \frac{25}{12 \times 6 \times 0.438} = 0.793$$

Using simplified expression for δ, Eq. (338), on the other hand gives

$$U_1/U_2 = 0.793 \times 0.438/0.328 = 1.058$$

Design Problem 25: In calculating the shift of the neutral axis for a curved member of rectangular cross-section illustrate graphically the range of applicability of the simplified formula. Present all results in nondimensional form using δ/R and R/h as parameters.

Solution: Dividing both sides of Eq. (334) by R and introducing $\chi = R/h$ gives

$$\frac{\delta}{R} = 1 - \frac{1}{\chi \log_e \left(\dfrac{2\chi + 1}{2\chi - 1} \right)}$$

With similar rearrangement and substitution, Eq. (338), yields

$$\frac{\delta}{R} = \frac{1}{1 + 12\chi^2}$$

The above nondimensional expressions are illustrated graphically in Fig. 12-5 for values of χ varying between 0.6 and 5.0.

Symbols for Chapter 12

A	Area of cross-section, in.^{2}
b	Width of rectangular section, in.
c	Distance of extreme fiber from central axis, in.
E	Modulus of elasticity, psi
e	Strain
$f, f(x), f(h)$	Functions of one variable
h	Depth of cross-section, in.

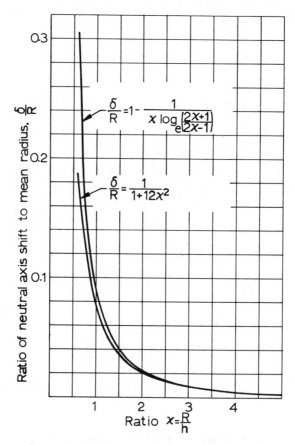

Fig. 12-5. Comparison of shift of neutral axis by exact and approximate formulas
for a rectangular cross-section

I	Moment of inertia, in.^{4}
M	Bending moment, lb-in.
N	Normal force, lb
P	External load, lb
R	Radius to center of gravity, in.
R_n	Radius to neutral axis, in.
R_1	Radius to inner surface of curved member, in.
R_2	Radius to outer surface of curved member, in.

S	Stress, psi
S_b	Bending stress, psi
U, U_1, U_2	Elastic strain energy, lb-in.
W_q	Wilson and Quereau stress correction factor
x	Arbitrary distance, in.
y	Distance from neutral axis, in.
δ	Distance from neutral to central axis, in.
θ	Angle at which forces are considered, rad
ϱ	General symbol for radius of curvature, in.
ψ	Slope, rad
$\chi = R/h$	Ratio of radius of curvature to depth of section

CHAPTER THIRTEEN

Design of Thick Curved Members

Design Considerations

It was shown in Chapter 2 that early developments of curved beam theory date back some two hundred years ago. This is not surprising when it is realized the enormous variety of machines and structures in all branches of industry involving some kind of curved members. Hence the need for design methods and formulas for such load carrying members as proving rings, rolling-element bearing rings, machine frames, load hooks, chain links, shackles and curved brackets of every description.

The elementary formulas for straight beams, often applied to curved beam design, become more in error as the neutral axis of a sharply curved member shifts toward the center of curvature. The theory of neutral axis indicates that the shift is primarily a function of the degree of curvature and the depth of the cross-section. The effect of the sectional geometry appears to be of somewhat lesser significance.

The basic question which often arises, concerns the problem of differentiating between the so called thin and thick members. An often quoted rule of thumb states that when the mean radius of curvature increases to about ten times the depth of the cross-section, the member is considered thin. In these cases all the previously developed design formulas apply with negligible error.

The rule of "ten times the depth" is considered relatively flexible and should be viewed with particular regard to the accuracy of the calculations required and the knowledge of material allowables. In fact the rule can often be relaxed since the true mechanical properties of engineering materials are seldom known with an accuracy better than plus or minus ten percent. Also the available experimental evidence, described later in this chapter, suggests that the limiting ratio of R/h as low as 6 may be acceptable.

Strength of Curved Beams

In calculating bending stresses in essentially straight beams the usual engineering formula $S_b = M/Z$, gives satisfactory results. However, in dealing with curved machine members, characterized by sharp curvature, the stresses developed at the inner face of the beam may be substantially higher than those calculated by the usual theory of straight members.

The original Winkler-Bach formula, Eq. (1), which takes into account beam curvature is theoretically correct and relatively simple in form, but requires determining section property parameter m defined by Eq. (2). To simplify the design procedure Wilson and Quereau developed a correction factor. The numerical values of this factor are given in Table 13-1 for typical cross-sections encountered in curved-beam design, (Ref. 15). The factors are applicable to the calculations of the maximum stresses at the outer and inner faces.

Table 13-1. Wilson and Quereau Factors for Stresses in Curved Beams

| | | W_q | | |
	R/c	Inner Face	Outer Face	δ/R
	1.2	3.41	0.54	0.224
	1.4	2.40	0.60	0.151
	1.6	1.96	0.65	0.108
	1.8	1.75	0.68	0.084
	2.0	1.62	0.71	0.069
	3.0	1.33	0.79	0.030
	4.0	1.23	0.84	0.016
	6.0	1.14	0.89	0.0070
	8.0	1.10	0.91	0.0039
	10.0	1.08	0.93	0.0025

Table 13-1 (cont.). Wilson and Quereau Factors for Stresses in Curved Beams

R/c	W_q Inner Face	Outer Face	δ/R
1.2	2.89	0.57	0.305
1.4	2.13	0.63	0.204
1.6	1.79	0.67	0.149
1.8	1.63	0.70	0.112
2.0	1.52	0.73	0.090
3.0	1.30	0.81	0.041
4.0	1.20	0.85	0.021
6.0	1.12	0.90	0.0093
8.0	1.09	0.92	0.0052
10.0	1.07	0.94	0.0033

R/c	W_q Inner Face	Outer Face	δ/R
1.2	3.01	0.54	0.336
1.4	2.18	0.60	0.229
1.6	1.87	0.65	0.168
1.8	1.69	0.68	0.128
2.0	1.58	0.71	0.102
3.0	1.33	0.80	0.046
4.0	1.23	0.84	0.024
6.0	1.13	0.88	0.011
8.0	1.10	0.91	0.0060
10.0	1.08	0.93	0.0039

4.

R/c	W_q Inner Face	Outer Face	δ/R
1.2	3.09	0.56	0.336
1.4	2.25	0.62	0.229
1.6	1.91	0.66	0.168
1.8	1.73	0.70	0.128
2.0	1.61	0.73	0.102
3.0	1.37	0.81	0.046
4.0	1.26	0.86	0.024
6.0	1.17	0.91	0.011
8.0	1.13	0.94	0.0060
10.0	1.11	0.95	0.0039

Table 13-1 (cont.). Wilson and Quereau Factors for Stresses in Curved Beams

R/c	W_q Inner Face	W_q Outer Face	δ/R
1.2	3.14	0.52	0.352
1.4	2.29	0.54	0.243
1.6	1.93	0.62	0.179
1.8	1.74	0.65	0.138
2.0	1.61	0.68	0.110
3.0	1.34	0.76	0.050
4.0	1.24	0.82	0.028
6.0	1.15	0.87	0.012
8.0	1.12	0.91	0.0060
10.0	1.10	0.93	0.0039

R/c	W_q Inner Face	W_q Outer Face	δ/R
1.2	3.26	0.44	0.361
1.4	2.39	0.50	0.251
1.6	1.99	0.54	0.186
1.8	1.78	0.57	0.144
2.0	1.66	0.60	0.116
3.0	1.37	0.70	0.052
4.0	1.27	0.75	0.029
6.0	1.16	0.82	0.013
8.0	1.12	0.86	0.0060
10.0	1.09	0.88	0.0039

R/c	W_q Inner Face	W_q Outer Face	δ/R
1.2	3.63	0.58	0.418
1.4	2.54	0.63	0.299
1.6	2.14	0.67	0.229
1.8	1.89	0.70	0.183
2.0	1.73	0.72	0.149
3.0	1.41	0.79	0.069
4.0	1.29	0.83	0.040
6.0	1.18	0.88	0.018
8.0	1.13	0.91	0.010
10.0	1.10	0.92	0.0065

Table 13-1 (cont.). Wilson and Quereau Factors for Stresses in Curved Beams

R/c	W_q Inner Face	W_q Outer Face	δ/R
1.2	3.55	0.67	0.409
1.4	2.48	0.72	0.292
1.6	2.07	0.76	0.224
1.8	1.83	0.78	0.178
2.0	1.69	0.80	0.144
3.0	1.38	0.86	0.067
4.0	1.26	0.89	0.038
6.0	1.15	0.92	0.018
8.0	1.10	0.94	0.010
10.0	1.08	0.95	0.0065

R/c	W_q Inner Face	W_q Outer Face	δ/R
1.2	2.52	0.67	0.408
1.4	1.90	0.71	0.285
1.6	1.63	0.75	0.208
1.8	1.50	0.77	0.160
2.0	1.41	0.79	0.127
3.0	1.23	0.86	0.058
4.0	1.16	0.89	0.030
6.0	1.10	0.92	0.013
8.0	1.07	0.94	0.0076
10.0	1.05	0.95	0.0048

R/c	W_q Inner Face	W_q Outer Face	δ/R
1.2	3.28	0.58	0.269
1.4	2.31	0.64	0.182
1.6	1.89	0.68	0.134
1.8	1.70	0.71	0.104
2.0	1.57	0.73	0.083
3.0	1.31	0.81	0.038
4.0	1.21	0.85	0.020
6.0	1.13	0.90	0.0087
8.0	1.10	0.92	0.0049
10.0	1.07	0.93	0.0031

Table 13-1 (cont.). Wilson and Quereau Factors for Stresses in Curved Beams

| | W_q | | |
R/c	Inner Face	Outer Face	δ/R
1.2	2.63	0.68	0.399
1.4	1.97	0.73	0.280
1.6	1.66	0.76	0.205
1.8	1.51	0.78	0.159
2.0	1.43	0.80	0.127
3.0	1.23	0.86	0.058
4.0	1.15	0.89	0.031
6.0	1.09	0.92	0.014
8.0	1.07	0.94	0.0076
10.0	1.06	0.95	0.0048

Approximate Formula for Maximum Bending Strength of Curved Members

The values of stress correction factor W_q, given in Table 13-1, help to make a rapid estimate of the bending stress at the inner and outer faces of the curved member. There is also a uniform tension or compression across the section of the beam analyzed for bending and this is why Eq. (343) contains the term P/A. In this equation the term P/A, is shown as positive. In the actual case, care should be taken to interpret the stresses due to bending and direct loading, respectively. For instance the tensile stress due to the direct loading must be added directly to the tensile stress due to bending and subtracted from the compressive stress due to bending. Of course the reverse is also true and the compressive stress due to the direct loading should be subtracted from the tensile stress due to bending and added directly to the compressive stress resulting from bending.

Examination of Table 13-1 indicates that despite rather striking differences in the cross-sectional geometry of the curved members considered, the values of correction factors W_q do not differ excessively for the R/c ratios higher than about 2. Plotting the available W_q factors from Table 13-1, for the case of the maximum bending stress at the inner surface of the beam, shows that the entire range of W_q values may be represented by the following approximate equation

$$W_q = 0.35 \left(\frac{3R^2 - Rc - c^2}{R^2 - Rc} \right)^* \qquad (345)$$

* For meaning of symbols and dimensional units involved for this and other equations in this chapter see material at end of chapter.

For instance taking $R/c = 4$, the calculated factor from Eq. (345), is

$$W_q = 0.35 \frac{3 \times 16c^2 - 4c^2 - c^2}{16c^2 - 4c^2} = 0.35 \times \frac{43}{12} = 1.26$$

Looking over the corresponding values in Table 13-1, shows that the correction factors vary between 1.15 and 1.29. This accuracy may be acceptable for the preliminary design calculations. The design curve based on the approximate formula, Eq. (345) is shown in Fig. 13-1. For more exact values Table 13-1 is recommended.

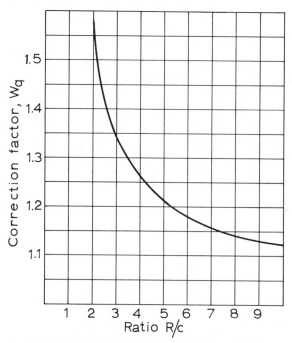

Fig. 13-1. Approximate correction factor for maximum bending strength of curved members

It should be noted that the correction factor W_q is used only to calculate maximum stresses provided there are no abrupt changes in cross-section. For unusually wide and thin webbed members this factor must be applied with caution because of the tendency of thin sections to fail by buckling due to compression or fracture at the free edges due to tension. This is of special importance when the curved member is made of relatively brittle material. To prevent such failures, often welded or riveted flanges are provided.

In the H-, I- and T-type common structural shapes the maximum radial stresses occur usually at the junction between the web and the inner flange. In simple theory however, the radial stresses are neglected as being considerably smaller than the stresses acting normal to the beam cross-section.

Curved Beam Under Horizontal End Load

In analyzing deflection of a relatively thick curved member Eq. (49) should be employed which takes into account the total elastic strain energy due to bending, normal, and shear stresses. The moment of inertia is replaced by the term, $AR\delta$, which signifies the effect of the shift of the neutral axis and ξ^*, denoting the shear distribution factor, enters into considerations whenever dealing with relatively thick members. The mean radius of curvature R, is measured to the center of gravity of the section as before. M, N and Q will denote the bending moment, normal force, and shear force, respectively, at an arbitrary section defined by θ. The expressions for the normal and shear forces follow directly from the geometry depicted in Figs. 13-2 and 13-3. The procedure involves simply considering the equilibrium at a section defined by θ, by transferring the end load to the center of gravity of that section and by resolving the transferred load into the radial and tangential components as shown. Here the tangential component becomes the nor-

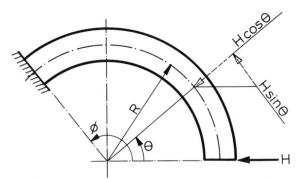

Fig. 13-2. Curved beam under horizontal end load

---

* The error caused by the use of the shear distribution factor in strain calculations, instead of a more accurate form factor (Ref. 11), will usually be found negligible and on the conservative side.

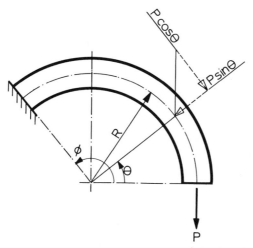

Fig. 13-3. Curved beam under vertical end load

mal force and the radial component is featured as a shear force on the cross-section under consideration.

Following this procedure and adhering to a consistent sign convention the bending moment and the relevant forces for the case illustrated in Fig. 13-2, are

$$M = -HR \sin \theta \qquad (346)$$

$$N = H \sin \theta \qquad (347)$$

$$Q = H \cos \theta \qquad (348)$$

The Castigliano equation expressing the deflection of the end of the beam in the direction of H is

$$X = R \int_0^\phi \left(\frac{M}{AER\delta} \frac{\partial M}{\partial H} + \frac{N}{AE} \frac{\partial N}{\partial H} + \frac{\xi Q}{AG} \frac{\partial Q}{\partial H} - \frac{M}{AER} \frac{\partial N}{\partial H} - \frac{N}{AER} \frac{\partial M}{\partial H} \right) d\theta \qquad (349)$$

With the aid of the elementary process of differentiating Eqs. (346) through (348) with respect to H we obtain the partial derivative terms needed for the solution of Eq. (349). This gives

$$\frac{\partial M}{\partial H} = -R \sin \theta$$

$$\frac{\partial N}{\partial H} = \sin \theta$$

$$\frac{\partial Q}{\partial H} = \cos \theta$$

Substituting the above terms together with the Eqs. (346), (347) and (348), into Eq. (349) and integrating with respect to θ yields the following general design formula for the deflection under horizontal load:

$$X = \frac{HR}{4AE}(2\phi - \sin 2\phi)\left(\frac{R}{\delta} + 3\right) + \frac{\xi HR}{4AG}(2\phi + \sin 2\phi)$$

$$(350)$$

Employing rigorous mathematical methods of the theory of elasticity Timoshenko and Goodier derived the deflection formula for a thick curved bar subtending 90 degrees and supporting a horizontal load, (Ref. 1). Utilizing notation adopted in this book their formula for a bar of rectangular cross-section may be written as follows

$$X = \frac{\pi H}{bE}\frac{(1 + 4\chi^2)}{\left[(1 + 4\chi^2)\log_e\left(\frac{2\chi + 1}{2\chi - 1}\right) - 4\chi\right]}$$

$$(351)$$

In Eq. (351), χ denotes the ratio of the radius of curvature to the depth of the cross-section. Substituting $\phi = \pi/2$, $G = 0.4E$, and $\xi = 3/2$, for a rectangular cross-section, into Eq. (350), gives

$$X = \frac{\pi HR}{16AE\delta}(4R + 27\delta)$$

$$(352)$$

Here δ denotes the shift of the neutral axis. For a rectangular cross-section this value is given by Eq. 334. For several other shapes, δ can be calculated with the aid of Table 13-1. Equations (351) and (352) are in close agreement. For instance, taking a rectangular cross-section with $b = 1$ in. and $h = 2$ in., and the mean radius of curvature $R = 3$ in., these equations give $10.7\ \pi H/E$ and $11.2\ \pi H/E$, respectively.

When δ becomes very small, Eq. (352) gives the design formula for a thin curved member subtending 90 degrees and carrying a horizontal concentrated load.

$$X = \frac{\pi HR^3}{4EI}$$

$$(353)$$

Equation (353) follows directly from the previously obtained formula, Eq. (86), when $\phi = \pi/2$, is substituted.

Curved Beam Under Vertical End Load

When a curved beam of a relatively deep cross-section, Fig. 13-3, is subjected to a vertical concentrated load applied at the free end, the moment and force equations may be written as follows:

$$M = -PR \left(1 - \cos \theta\right)$$

$$N = -P \cos \theta$$

$$Q = P \sin \theta$$

The corresponding partial derivatives with respect to P are

$$\frac{\partial M}{\partial P} = -R \left(1 - \cos \theta\right)$$

$$\frac{\partial N}{\partial P} = -\cos \theta$$

$$\frac{\partial Q}{\partial P} = \sin \theta$$

The Castigliano equation for deriving the design formula for the vertical deflection, Fig. 13-3, is of the same general form as Eq. (349) except the partial derivative terms are now with respect to P instead of H and the moment and force equations are now expressed in terms of P, as shown. Hence utilizing the above relations, the deflection formula becomes

$$Y = \frac{PR}{4AE} \left[2\phi \left(\frac{3R}{\delta} - 1\right) + \left(\sin 2\phi - 8 \sin \phi\right)\left(\frac{R}{\delta} - 1\right)\right]$$
$$+ \frac{PR\xi}{4AG} \left(2\phi - \sin 2\phi\right) \tag{354}$$

For a curved member subtending 90 degrees, Eq. (354), gives

$$Y = \frac{PR}{4AE} \left[\frac{R}{\delta} \left(3\pi - 8\right) - \left(\pi - 8\right)\right] + \frac{PR\pi\xi}{4AG} \tag{355}$$

When δ becomes very small, all terms multiplied by δ in the above formula, may be ignored. Also making $AR\delta = I$, Eq. (355), yields

$$Y = \left(\frac{3\pi - 8}{4}\right)\frac{PR^3}{EI} \tag{356}$$

It may be noted that Eq. (356) follows from Eq. (77), for thin arcuate members when $\phi = \pi/2$ is substituted. For a specific case of a thick curved bar of rectangular cross-section shear distribution factor $\xi = 3/2$. For most metallic materials $G = 0.4E$. Substituting these numerical values, Eq. (355) transforms into

$$Y = \frac{PR}{16AE\delta} \left[4R (3\pi - 8) + (11\pi + 32) \delta \right] \qquad (357)$$

Formulas given by Eqs. (352) and (357) may be used conveniently to compare flexibility of a quarter circle thick bar when loaded horizontally and vertically. For instance dividing Eq. (352) by Eq. (357), and making $H = P$, gives

$$\frac{X}{Y} = \frac{4\pi + \dfrac{27\pi\delta}{R}}{4 (3\pi - 8) + \dfrac{(11\pi + 32) \delta}{R}}$$

When δ approaches zero $X/Y = \pi/(3\pi - 8) = 2.21$. For the maximum value of δ/R, for a rectangular section Table 13-1, this ratio reduces to $X/Y = 1.48$. This is a significant reduction and should be considered when designing the type of members shown in Figs. 13-2 and 13-3, for rigidity.

Thick Ring in Diametral Compression

A thick elastic ring working as a compression member is an important machine element which has varied applications. It occurs as a bearing ring, rim of heavy gear, proving ring or a rigid frame. In such applications the maximum stress may be limited by the allowable deflection and therefore there is a need for a convenient design formula, describing the relationship between the applied load and deflection.

A typical thick, elastic ring, subject to diametral compression, is shown in Fig. 13-4. Consider the equilibrium of one quarter of the ring fixed at the point of application of compressive load P, and carrying fixing moment M_f, reaction $P/2$, and a fictitious load $H/2$ acting at the lower end of the quadrant where $\theta = 0$. Because of symmetry it is necessary to analyze only one quarter of the ring. For an arbitrary section of the ring, such as defined by θ, the following relations hold

$$M = \frac{PR}{2} (1 - \cos \theta) + \frac{HR}{2} \sin \theta - M_f \qquad (358)$$

$$N = \frac{H}{2} \sin \theta - \frac{P}{2} \cos \theta \qquad (359)$$

$$Q = \frac{P}{2} \sin \theta + \frac{H}{2} \cos \theta \qquad (360)$$

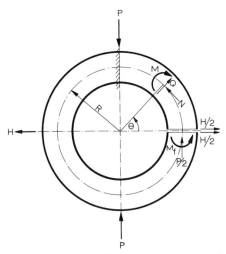

Fig. 13-4. Thick ring in diametral compression

The sign convention adopted in the above relations is consistent with the interpretation of the general expression for strain energy for a curved bar of a relatively deep cross-section given by Timoshenko, (Ref. 17).

The statically indeterminate fixing couple M_f can be found with the aid of the Castigliano theorem, Eq. (50), on the assumption that bending slope $\psi = 0$ when $\theta = 0$. This gives

$$\int_0^{\pi/2} (M - N\delta) \frac{\partial M}{\partial M_f} \, d\theta = 0 \qquad (361)$$

Utilizing Eqs. (358) and (359), integrating Eq. (361) and solving for M_f, yields

$$M_f = \frac{PR}{2} \left(1 - \frac{2}{\pi} + \frac{2\delta}{\pi R} \right) \qquad (362)$$

When δ becomes infinitely small, Eq. (362), reduces to Eq. (231), derived for a thin ring in diametral compression. For the maximum ratio $\delta/R = 0.418$, as given in Table 13-1, Eq. (362), gives $M_f = 0.315\,PR$. For a thin ring, Eq. (231), gives $M_f = 0.182\,PR$. Therefore the effect of the increasing shift of the neutral axis is to increase the magnitude of the redundant quantity M_f.

Once the redundant couple M_f is calculated the bending moment expression, Eq. (358) becomes

$$M = \frac{P}{2\pi} (2R - 2\delta - \pi R \cos \theta) + \frac{HR}{2} \sin \theta \qquad (363)$$

The vertical deflection is now obtained from the following equation:

$$Y = 4R \int_0^{\pi/2} \left(\frac{M}{AER\delta} \frac{\partial M}{\partial P} + \frac{N}{AE} \frac{\partial N}{\partial P} + \frac{\xi Q}{AG} \frac{\partial Q}{\partial P} - \frac{M}{AER} \frac{\partial N}{\partial P} \right.$$

$$\left. - \frac{N}{AER} \frac{\partial M}{\partial P} \right) d\theta \tag{364}$$

In using Eqs. (359), (360) and (363) with Eq. (364), the fictitious term H is omitted. Integrating, Eq. (364) and substituting the limits gives

$$Y = \frac{P(\pi^2 R^2 - 8R^2 + 8\delta^2)}{4\pi AE\delta} + \frac{PR\pi}{4AE} + \frac{PR\pi\xi}{4AG} - \frac{2P(\pi^2 R + 8\delta - 8R)}{4\pi AE} \tag{365}$$

By analogy to Eq. (364) the horizontal deflection is

$$X = 4R \int_0^{\pi/2} \left(\frac{M}{AER\delta} \frac{\partial M}{\partial H} + \frac{N}{AE} \frac{\partial N}{\partial H} + \frac{\xi Q}{AG} \frac{\partial Q}{\partial H} - \frac{M}{AER} \frac{\partial N}{\partial H} \right.$$

$$\left. - \frac{N}{AER} \frac{\partial M}{\partial H} \right) d\theta \tag{366}$$

Here H denotes a fictitious force acting along the horizontal center line as shown in Fig. 13-4. Evaluating the partial derivatives with respect to H from Eqs. (359), (360), and (363), integrating Eq. (366), and finally making the fictitious force H equal to zero yields

$$X = \frac{P(4R - \pi R - 4\delta)}{2\pi AE} \left(\frac{R}{\delta} - 1 \right) + \frac{PR\xi}{2AG} \tag{367}$$

Equations (365) and (367) may be simplified by ignoring the terms involving δ^2 as being relatively small. This gives

$$Y = \frac{PR}{4A} \left[\frac{(\pi^2 - 8)(R - 2\delta)}{\pi E\delta} + \pi \left(\frac{1}{E} + \frac{\xi}{G} \right) \right] \tag{368}$$

$$X = \frac{PR}{2A} \left[\frac{(4 - \pi)(R - \delta)}{\pi E\delta} - \frac{4}{\pi E} + \frac{\xi}{G} \right] \tag{369}$$

For the usual proportions of thick rings of rectangular cross-section, and the usual ratio of G/E for metals, Eqs. (365) and (367) may be expressed in terms of χ, which is the ratio of mean radius of ring to depth of cross-section (Ref. 31).

$$Y = \frac{P\chi}{bE} [1.788\chi^2 + 3.091 - 0.637/(1 + 12\chi^2)] \tag{370}$$

and

$$X = \frac{P\chi}{bE} \, [1.644\chi^2 + 0.926 + 0.637/(1 + 12\chi^2)] \qquad (371)$$

Tests on Thick Rings

Despite the importance of the load-deflection characteristics of a relatively thick ring of rectangular cross-section in design of machine parts such as proving rings, hollow rollers, rolling-element bearing rings, and similar components, deflection tests appear to have been rather few.

Goodenough and Moore (Ref. 2) conducted deflection tests on a steel ring having the ratio of outer to inner ring diameter of 1.3. Their calculated deflections based on the contribution of bending and direct stresses were in agreement with experimental results. Some years later similar tests were made at Kiev Polytechnicum in Russia in which the ratio went only as high as 1.2, and the results were never published *. Pippard and Miller (Ref. 5) made calculations and tests for a circular ring having ratio of 1.22. They considered strain energy due to bending, shear, and normal stresses but ignored the allowance for shear distribution and displacement of neutral axis. The tests were found to be in agreement with computations.

The calculation and tests made on thick circular rings at London University covered the ratios ranging from 1.3 to 1.92, (Ref. 29). The rings were tested as compressed by two forces along the vertical diameter, Fig. 13-4, and the deflection was recorded at the inner surface. In calculating the deflection the principle of Castigliano and expressions for strain energy due to bending, normal, and shear stresses were employed. The derivations included shear distribution factor and shift of the neutral axis effects on the deflection, and the required modulus of elasticity of the material was determined experimentally.

In the late 30's Maulbetsch and Nelson at the University of Michigan obtained an exact solution for the stresses and displacement of a thick ring in diametral compression but their work was not published †.

A comparison of some of the experimental and theoretical results obtained on thick rings is shown in Fig. 13-5. The deflection ratios are shown as a function of outer to inner diameter ratio. Here Δ

---

* Communicated to the author by letter (1955) from S. Timoshenko.
† Communicated to the author by letter (1959) from J. L. Maulbetsch.

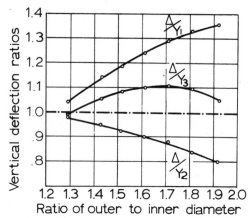

Fig. 13-5. Comparison of test and theory for a thick ring in compression

denotes the vertical deflection measured at the inner surface of the ring loaded in diametral compression as shown in Fig. 13-4. The theoretical value of the deflection, obtained with the aid of thin ring theory, Eq. (51), is denoted by Y_1. The corresponding deflection Y_2 is derived from the curved beam theory expressed for this case by Eq. (364). Symbol Y_3, denotes the vertical deflection obtained by Nelson and is based on the more rigorous solution of the theory of elasticity.

In essence, the solution of many deflection problems encountered in curved member design, is approximate in nature, though quite satisfactory for most practical purposes. The test results, obtained by various investigators suggest that the thin ring theory does not give too serious errors provided the ratio of outer to inner ring diameter does not exceed 1.2. In terms of the often employed terminology in design this ratio is equivalent to about $R/h = 5.5$. It is also interesting to note from Fig. 13-5, that extrapolating the three curves to about this ratio there appears to be little difference between the values derived from the thin ring and curved beam theories. However the trend of the curve derived with the aid of the theory of elasticity seems to indicate that for relatively thin rings, where naturally the displacements are larger, the solution of the theory of elasticity may become less accurate.

In the same series of tests (Ref. 29) it has been found that the agreement between the theory and measurements, in a horizontal sense, was better (Ref. 31). The discrepancy between the curved beam theory and tests in a vertical deflection is more serious and is probably due to the effect of radial strain in the line of action of the compressive load. Further research work in this area may be needed in specific engineering applications.

Numerical Examples

Design Problem 26: A machine clamp subtending 180 deg is made in structural steel having a yield strength of 60,000 psi and a modulus of elasticity $E = 30 \times 10^6$ psi. Assuming the dimensions shown in Fig. 13-6, calculate the limiting normal load P, so as not to exceed

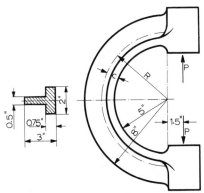

Fig. 13-6. Half-circle clamp

the elastic limit of the material. Find the corresponding maximum spread of the clamp jaws ignoring the deflection of the straight portions of the clamp, employing curved beam theory. Take $G = 0.4E$, and shear distribution factor, $\xi = 1.5$. Check the deflection using the thin member formula.

Solution: The stress and deflection formulas applicable to this case are given by Eqs. (343) and (355), respectively. To find the radius of curvature to the gravity axis, first determine the position of the center of gravity

$$c = \frac{3 \times 0.5 \times 1.5 + 0.75 \,(2 - 0.5) \times 0.375}{3 \times 0.5 + 0.75 \,(2 - 0.5)} = 1.02 \text{ in.}$$

Hence

$$R = 5 + 1.02 = 6.02 \text{ in.}$$

and

$$R/c = 5.9$$

For case 7, Table 13-1, by interpolation

$$\delta/R = 0.019$$

and

$$\delta = 0.019 \times 6.02 = 0.1145 \text{ in.}$$

The maximum bending moment about the central axis, Fig. 13-6, is

$$M = (6.02 + 1.5) P = 7.52P$$

Since the maximum stress will be at the inner surface of the clamp, interpolating, $R/c = 5.9$, and the stress correction factor becomes

$$W_q = 1.186$$

The moment of inertia of the clamp cross-section about the gravity axis is

$$I = \frac{2 \times 0.75^3}{12} + \frac{0.5 \times (3 - 0.75)^3}{12} + (2 \times 0.75) (1.02 - 0.375)^2$$

$$+ (3 - 0.75) \times 0.5 [3 - 1.02 - (3 - 0.75)/2]^2$$

$$I = 0.070 + 0.476 + 0.624 + 0.810 = 1.995 \text{ in.}^4$$

The cross-sectional area is

$$A = 3 \times 0.5 + 0.75 (2 - 0.5) = 2.625 \text{ in.}^2$$

Hence substituting the numerical data in Eq. (343), gives

$$60,000 = 1.186 \left[\frac{P}{2.625} + \frac{7.52P \times 1.02}{1.995} \right]$$

from which

$$P = 12,100 \text{ lb}$$

The total spread of the clamp jaws under this load is equal to twice the deflection given by Eq. (355) since the clamp can be represented as two identical curved beams joined at the horizontal axis x-x. Substituting the relevant numerical values gives

$$Y = \frac{12,100 \times 6.02}{2 \times 2.625 \times 30 \times 10^6} \left[\frac{6.02}{0.1145} (3\pi - 8) - (\pi - 8) \right]$$

$$+ \frac{12,100 \times 6.02 \times \pi \times 1.5}{2 \times 2.625 \times 0.4 \times 30 \times 10^6}$$

$$Y = 0.042 \text{ in.}$$

It is seen that the contribution of the shear and direct stresses in this particular case is relatively small. Using thin member formula, Eq. (356), which considers bending only and ignores the effect of the shift of the neutral axis, yields

$$Y = \frac{3\pi - 8}{2} \times \frac{12{,}100 \times 6.02^3}{1.995 \times 30 \times 10^6}$$

$$Y = 0.031 \text{ in.}$$

Design Problem 27: A rigid end support hook, Fig. 13-7, has elliptical cross-section and subtends 180 deg. Find the limiting load on the hook if the material is cast-iron having tensile strength of 25,000 psi. Calculate the theoretical loading capacity utilizing Wilson and Quereau stress correction factor and the design curve method.

Solution: With reference to Fig. 13-7, the ratio $R/c = 5/2 = 2.5$. From Table 13-1, case 1, Wilson and Quereau factor for the inner face of the hook is $W_q = 1.48$, by straight line interpolation. The maximum bending moment at the built-in end of the hook is $M = 2PR$. Hence substituting the moment in Eq. (343) and solving for P gives

$$P = \frac{S}{W_q\left(\dfrac{1}{A} + \dfrac{2Rc}{I}\right)}$$

The moment of inertia for the elliptical cross-section shown in Fig. 13-7, is $I = 2\pi$ in.⁴, and the corresponding cross-sectional area $A = 2\pi$ in.². Hence substituting the numerical data in the above equation the required loading capacity is

$$P = \frac{25{,}000}{1.48\left(\dfrac{1}{2\pi} + \dfrac{2 \times 2 \times 5}{2\pi}\right)}$$

$$P = 5070 \text{ lb}$$

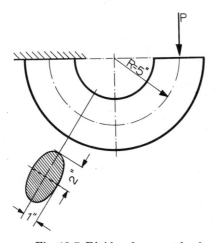

Fig. 13-7. Rigid end support hook

Using the approximate correction factor corresponding to $R/c = 2.5$, from the curve in Fig. 13-1, we get

$$W_q = 1.41$$

Hence the approximate loading capacity is

$$P = 5070 \times 1.48/1.41$$

$$P = 5320 \text{ lb}$$

The difference between the two values of loading is only about 5 percent.

Design problem 28: A short pipe element dimensioned in Fig. 13-8,

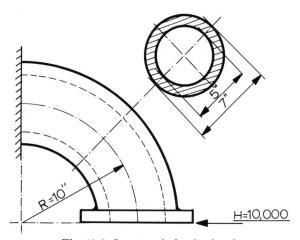

Fig. 13-8. Quarter-circle pipe bend

is rigidly fixed at one end and carries a horizontal thrust load of $H = 10,000$ lb. Assuming that the material is cast steel with the yield tensile strength of 35,000 psi and $E = 30 \times 10^6$ psi, find the approximate margin of safety for the design. Calculate the deflection under horizontal thrust using curved beam and thin ring theories. Assume shear distribution factor to be equal to one, and modulus of rigidity $G = 0.4E$.

Solution: For a ratio $R/c = 10/3.5 = 2.86$, the approximate correction factor for the maximum stress $W_q = 1.355$. The maximum bending moment is $M = HR = 100,000$ lb-in. The moment of inertia may be calculated using the formula

$$I = \pi r^3 t = \pi \times 3^3 \times 1 = 27\pi \text{ in.}^4$$

The corresponding cross-sectional area is

$$A = 2\pi r t = 6\pi \text{ in.}^2$$

Hence, using Eq. (343), gives

$$S = 1.355 \left(\frac{10,000}{6\pi} + \frac{100,000 \times 3.5}{27\pi} \right)$$

$$S = 6350 \text{ psi}$$

The margin of safety is therefore

$$\frac{35,000}{6350} - 1 = 4.5$$

Substituting $\phi = \pi/2$, $G = 0.4E$ and $\xi = 1$, in Eq. (350), yields

$$X = \frac{\pi H R}{4 A E \delta} (R + 5.5\delta)$$

Making a conservative estimate for δ/R based on Table 13-1, cases 1 and 10, gives $\delta = 0.04 \times 10 = 0.4$ in. Hence substituting the relevant data in the above deflection formula, yields

$$X = \frac{\pi \times 10,000 \times 10}{4 \times 6\pi \times 0.4 \times 30 \times 10^6} (10 + 5.5 \times 0.4)$$

$$X = 0.0042 \text{ in.}$$

Utilizing Eq. (353) for a thin curved member, gives

$$X = \frac{\pi \times 10,000 \times 10^3}{4 \times 30 \times 10^6 \times 27\pi}$$

$$X = 0.0031 \text{ in.}$$

Design Problem 29: A heavy duty proving ring, such as that shown in Fig. 13-4, carries a compressive diametral load $P = 50,000$ lb. The material is good quality structural steel having the yield strength and the modulus of elasticity 100,000 psi and 30×10^6 psi respectively. The outer and inner ring diameters are 20 in. and 12 in. The ring cross-section is rectangular with the depth and width of 4 in. and 2 in. respectively. Calculate the change in vertical and horizontal diameters under the maximum working load.

Solution: The ratio of mean radius to depth of cross section is $\chi = R/h = 8/4 = 2$. Hence, utilizing Eqs. (370) and (371) gives

$$Y = \frac{50,000 \times 2}{2 \times 30 \times 10^6} \left[1.788 \times 4 + 3.091 - 0.637/(1 + 12 \times 4) \right]$$

$Y = 0.0170$ in.

and

$$X = \frac{50,000 \times 2}{2 \times 30 \times 10^6} \left[1.644 \times 4 + 0.926 + 0.637/(1 + 12 \times 4) \right]$$

$X = 0.0125$ in.

Design Problem 30: In manufacturing rolling-element bearing rings or housings, out-of-round components are often assembled inducing bending stresses. The parameter required for the calculation of such stresses is usually called out-of-roundness defined as the sum of the vertical and horizontal change of ring diameter. Utilizing this concept obtain: (1) a simplified formula for out-of-roundness in a bearing ring of rectangular cross-section and (2) the maximum bending stress as a function of out-of-roundness, Wilson-Quereau stress correction factor, modulus of elasticity, mean radius of the ring and the depth of ring cross-section.

Solution: Denoting out-of-roundness by Ω, and utilizing Eqs. (370) and (371), gives

$$\Omega = Y + X$$

$$\Omega = \frac{P\chi}{bE} (3.43\chi^2 + 4.02)$$

The maximum bending moment in a circular ring in diametral compression follows from Eq. (363).

$$M = \frac{P (R - \delta)}{\pi}$$

From Eq. (343), the corrected bending stress for a rectangular cross-section is

$$S_b = 6M \, W_q/bh^2$$

Eliminating bending moment and expressing the stress in terms of out-of-roundness Ω, yields

$$S_b = \frac{W_q E \Omega}{h (1.788\chi^2 + 2.090)}$$

In the above formula very small terms have been ignored.

Symbols for Chapter 13

A	Area of cross-section, in.^{2}
b	Width of section, in.
c	Distance from inner surface to central axis, in.
E	Modulus of elasticity, psi
G	Modulus of rigidity, psi
H	Horizontal load, lb
h	Depth of cross-section, in.
I	Moment of inertia, in.^{4}
M	Bending moment, lb-in.
M_f	Fixing moment, lb-in.
m	Section property in Winkler-Bach formula
N	Normal force, lb
P	Vertical load, lb
Q	Transverse shearing force, lb
R	Radius to center of gravity, in.
r	Mean radius of pipe, in.
S	Stress, psi
S_b	Bending stress, psi
t	Wall thickness, in.
W_q	Wilson and Quereau stress correction factor
X	Horizontal deflection, in.
Y	Vertical deflection, in.
Y_1	Vertical deflection, thin ring theory, in.
Y_2	Vertical deflection, curved beam theory, in.
Y_3	Vertical deflection, theory of elasticity, in.
Z	Section modulus, in.^{3}
Δ	Deflection by test, in.

δ	Distance from neutral to central axis, in.
θ	Angle at which forces are considered, rad
ϕ	Angle subtended by curved member, rad
ψ	Slope, rad
ξ	Shear distribution factor
$\chi = R/h$	Ratio of radius of curvature to depth of section
Ω	Out-of-roundness, in.

CHAPTER FOURTEEN

Curved Members With
Varying Cross-Sections

Engineering Considerations

In analyzing stress and rigidity characteristics of machines for rivetting, stitching, drilling, welding, and press and jig saw work as well as other standard mechanical operations, curved beam problems of various complexity are encountered. This chapter discusses the analysis of stress and flexural rigidity and presents the applications of well known principles to the design of curved members of nonuniform cross-sections. The material presented follows closely the original published work in this field by K. E. Lofgren (Ref. 32).

When curved beam sections are not uniform the mathematical process of obtaining deflections is greatly complicated and in many cases almost impossible to follow. Furthermore many machine frames are complicated by lateral ribs, mounting pads, special bosses and other irregular shapes so that the exact expressions for the variation of the moment of inertia along the gravity axis of the curved member are extremely difficult to develop. In such cases, of course, the designer is compelled to make various approximations to suit a particular need.

The techniques for calculating the deflections of curved members with varying cross-sections are usually considered to be beyond the scope of most college textbooks and engineering handbooks. This book adopts Lofgren's semigraphical and mathematical procedures as very useful tools in estimating flexural deflections for support frames and machine elements which depend on the maintaining of adequate rigidity for their satisfactory performance. The procedures presented are based on the expressions for strain energy due to bending and the theorem of Castigliano.

Lofgren's Semigraphical Method

The general form of a Castigliano equation for the deflection of a curved member with a varying cross-section is

$$Y = \frac{1}{E} \int_0^L \frac{M}{I_x} \frac{\partial M}{\partial P} \, ds \, * \tag{372}$$

Here I_x denotes the moment of inertia at an arbitrary point along the neutral axis of the curved member and is considered as a variable quantity. It is recalled here that in all previous derivations throughout this text, the moment of inertia was assumed to be constant and always remained outside the integral sign. The differential length is denoted by ds and the total stretched-out length by L.

Consider a typical curved member subjected to tensile forces P, as shown in Fig. 14-1. To find the deflection by the semigraphical method, divide the length along the neutral axis into an arbitrary number of equal parts. In this particular illustration we have 12 divisions. Next, calculate the moments of inertia for various cross-sections cor-

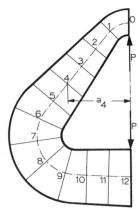

Fig. 14-1. Curved member with varying cross-section

* For meaning of symbols and dimensional units involved for this and other equations in this chapter see material at end of chapter.

responding to subdivisions, making certain that the relevant sections are perpendicular to the neutral axis. The main difficulty in performing such calculations lies in the fact that the position of the entire neutral axis must be known and the neutral axis cannot be found without having the centroid axis first. However, use of the design data, such as presented in Table 13-1, supplemented by good engineering judgement, should result in a set of values which will yield acceptable results.

For the case illustrated in Fig. 14-1, Eq. (372) can be restated as follows

$$Y = \frac{P}{E} \int_0^s \frac{a_x{}^2}{I_x} \, ds \qquad (373)$$

In Eq. (373), a_x represents a moment arm from line of action of load to an arbitrary point along the neutral axis. For instance the bending moment at station 4 is $M = Pa_4$. The partial derivative with respect

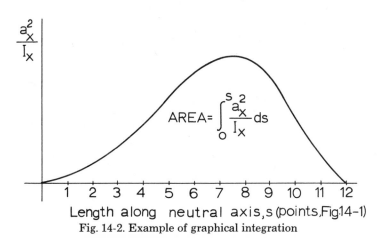

Fig. 14-2. Example of graphical integration

to P is $\partial M/\partial P = a_4$. Hence follows justification for the square of the moment arm a_x in Eq. (373). Since s is the distance along the neutral axis, the values $a_x{}^2/I_x$, can be plotted against s, as shown in Fig. 14-2. These values are obtained for each section, Fig. 14-1, and the area under the curve, Fig. 14-2, represents the integral from Eq. (373).

$$\int_0^s \frac{a_x{}^2}{I_x} \, ds$$

The area under this curve can be obtained directly by means of a planimeter or by a simple arithmetical summation method without the graphical construction. Once the summation is completed the deflection is obtained by multiplying this result by the term P/E. Note that the quantity under the integral sign Eq. (373) is inversely proportional to length so that the resultant dimension must be in length. This follows from dimensional analysis applied to the right of integral sign: in.^{2} $\times$ in./in.^{4}.

Lofgren's Parametric Method

The curved member of varying cross-section with partially straight length, shown in Fig. 14-3, represents a fairly common type of prob-

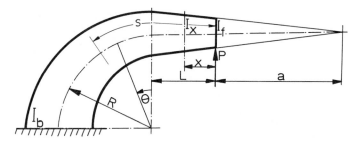

Fig. 14-3. Varying section curved member with straight length

lem occurring in machine design. The mathematical analysis in this case may be possible if a suitable relationship can be established between the moment of inertia I_x and the distance s, measured along the neutral axis curve. For the type of problem, depicted in Fig. 14-3, Lofgren proposes the following relation

$$\frac{I_b}{I_x} = \frac{(a + L + R)^2}{(a + x)^2} \tag{374}$$

If I_b and I_f shall denote the moments of inertia at the fixed and free ends of the curved member respectively, these terms as well as dimensions R and L can have many possible values. The length a is automatically defined by Eq. (374). To reduce the number of variables involved it is best to work through non-dimensional parameters such as I_b/I_f, L/R and a/R.

Assuming that the bending strain energy is predominant for the case shown in Fig. 14-3, the basic relation, Eq. (372), may be integrated separately along the straight and curved portions, respectively.

From Eq. (374), the general expression for the moment of inertia is

$$I_x = I_b \frac{(a+x)^2}{(a+L+R)^2}$$

The bending moment along the straight portion is simply $M = Px$ and the corresponding partial derivative $\partial M/\partial P = x$. Hence, using Eq. (372) gives

$$Y_1 = \frac{P(a+L+R)^2}{EI_b} \int_0^L \frac{x^2 \, dx}{(a+x)^2} \tag{375}$$

Integrating Eq. (375), yields

$$Y_1 = \frac{P(a+L+R)^2}{EI_b} \left[\frac{L(L+2a)}{a+L} + 2a \log_e \frac{a}{a+L} \right] \tag{376}$$

Substituting, for instance, $L = 2R$ and $a = 1.5R$ in the above equation gives

$$Y_1 = 6.38 \frac{PR^3}{EI_b} \tag{377}$$

The contribution of the curved portion to the deflection of the beam shown in Fig. 14-3 is obtained by integrating the following expression

$$Y_2 = \frac{PR}{EI_b} (a+L+R)^2 \int_0^{\pi/2} \frac{(L+R\sin\theta)^2}{(a+L+R\sin\theta)^2} \tag{378}$$

Integrating Eq. (378), and substituting $L = 2R$ and $a = 1.5R$ as before yields

$$Y_2 = 12.9 \frac{PR^3}{EI_b} \tag{379}$$

Hence the total deflection at the loaded end of the beam, Fig. 14-3, follows from Eqs. (377) and (379)

$$Y = Y_1 + Y_2$$

$$Y = \zeta \frac{PR^3}{EI_b} \tag{380}$$

In Eq. (380), ζ represents a function of nondimensional parameters I_b/I_f, L/R and a/R. The design curves for finding ζ have been calculated by Lofgren and are shown in Figs. 14-4 and 14-5. With the aid of these charts the deflection of various curved members of the type shown in Fig. 14-3 can easily be found.

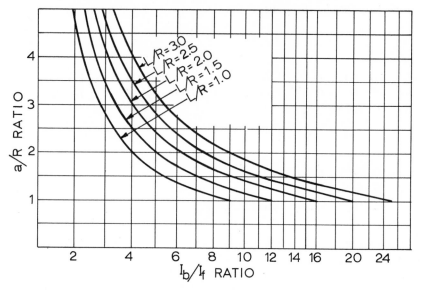

Fig. 14-4. Chart for finding a/R ratio (See Fig. 14-3)

Design of Varying Section Rings

Elastic rings of non-uniform cross-sections have varied applications ranging from small tool holders to large bulkheads. When variation in section is gradual the moment of inertia of the cross-section can be represented by a suitable continuous function. Denoting for instance the minimum and maximum moments of inertia by I_{min} and I_{max} respectively the following expression was proposed by Lee (Ref. 33)

$$I = I_{min}e^{n\theta} \qquad (381)$$

Here for a single-taper ring

$$n = \frac{1}{\pi} \log_e (I_{max}/I_{min}) \qquad (382)$$

In the above equations, e is the base of natural logarithm and θ denotes the angle at which the moments and forces are considered. A typical single taper ring is shown in Fig. 14-6. The ring is in equilibrium under a system of three parallel forces. If the ring is cut at the top, a system of forces N and M_f should be applied at $\theta = 0$, to simulate the constraining effect of the removed half of the ring as shown. The radius of curvature R, is measured to the neutral axis.

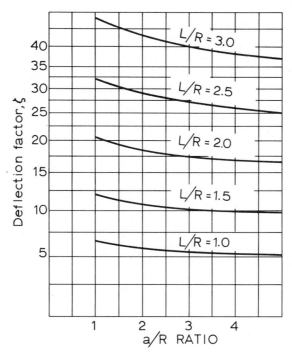

Fig. 14-5. Lofgren's deflection factor

Hence in accordance with the established sign convention the bending moment at an arbitrary angle θ, smaller than β, is

$$M = M_f - NR \, (1 - \cos \theta) \qquad (383)$$

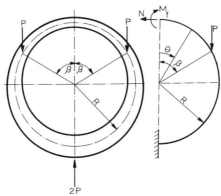

Fig. 14-6. Single taper ring under parallel forces

Note that in the above equation shear force at $\theta = 0$ is zero because of symmetry. The design equations for M_f and N can be now derived with the aid of the second principle of Castigliano which is equivalent to the method of the least work. Hence utilizing Eq. (47) and remembering that the moment of inertia is considered here as variable defined by Eq. (381), as well as using Eq. (383), yields the following expression for the elastic energy of bending for the portion of the ring subtended by angle β.

$$U_1 = \frac{R}{2EI_{\min}} \int_0^\beta [M_f - NR \, (1 - \cos \theta)]^2 \, e^{-n\theta} \, d\theta \qquad (384)$$

When $\theta > \beta$, the corresponding bending moment equation applies

$$M = M_f - NR \, (1 - \cos \theta) - PR \, (\sin \theta - \sin \beta) \qquad (385)$$

Hence, utilizing Eq. (385), the remaining part of the energy becomes

$$U_2 = \frac{R}{2EI_{\min}} \int_\beta^\pi [M_f - NR \, (1 - \cos \theta) - PR \, (\sin \theta - \sin \beta)]^2 \, e^{-n\theta} \, d\theta$$
$$(386)$$

Hence the total energy for one half of the ring is simply

$$U = U_1 + U_2 \qquad (387)$$

Since M_f and N are statically indeterminate, finding $\partial U / \partial M_f = 0$, and $\partial U / \partial N = 0$, from Eq. (387), gives two simultaneous equations for evaluating M_f and N. Using this general procedure Lee obtained the force and moment equations at $\theta = 0$ for the rings illustrated in Figs. 14-6 through 14-9. The equations given below have been abstracted from his original paper (Ref. 33).

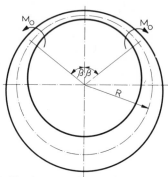

Fig. 14-7. Single taper ring under external couples

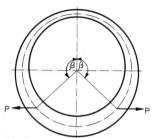

Fig. 14-8. Single taper ring under tensile forces

Case of Parallel Forces (Fig. 14-6)

$$N = \frac{P}{V_{12}} [nV_5 (V_3 \sin \beta - V_7) + V_{10} - V_6 \sin \beta] \qquad (388)$$

$$M_f = PR \left\{ (1 - nV_5) \frac{[nV_5 (V_3 \sin \beta - V_7) + V_{10} - V_6 \sin \beta]}{V_{12}} \right.$$
$$\left. - V_4 \sin \beta + V_9 \right\} \qquad (389)$$

Case of External Couples (Fig. 14-7)

$$N = \frac{M_o}{R} \left(\frac{n V_3 V_5 - V_6}{V_{12}} \right) \qquad (390)$$

$$M_f = \frac{M_o}{V_{12}} [(1 - nV_5) (nV_3V_5 - V_6) - V_4V_{12}] \qquad (391)$$

Case of Tensile Forces (Fig. 14-8)

$$N = \frac{P}{V_{12}} [nV_5 (V_3 \cos \beta - V_6) + V_{11} - V_6 \cos \beta] \qquad (392)$$

$$M_f = PR \left\{ (1 - nV_5) \frac{[nV_5 (V_3 \cos \beta - V_6) + V_{11} - V_6 \cos \beta]}{V_{12}} \right.$$
$$\left. + V_8 - V_4 \cos \beta \right\} \qquad (393)$$

Fig. 14-9. Single taper ring under internal pressure

Case of Internal Pressure (Fig. 14-9)

$$n_A = \frac{1}{\pi} \log_e (A_{max}/A_{min}) \tag{394}$$

$$M = M_f - (NR - qR^2) (1 - \cos \theta) \tag{395}$$

$$N = qR \left[1 - \frac{V_{13} (r_o/R)^2}{V_{12} + V_{14} (r_o/R)^2} \right] \tag{396}$$

$$M_f = qR^2 \left[\frac{V_{13} (nV_5 - 1) (r_o/R)^2}{V_{12} + V_{14} (r_o/R)^2} \right] \tag{397}$$

In equations for the case of varying section ring under internal pressure, r_o denotes radius of gyration of the smallest section in inches. A_{max} and A_{min} denote the maximum and minimum cross-section areas, respectively. Various load constants given in Eqs. (388) through (397) are summarized in Table 14-1.

Table 14-1. Force and Moment Coefficients for Varying Section Rings

$$V_1 = e^{-n\pi}$$

$$V_2 = e^{-n\beta}$$

$$V_3 = (V_2 - V_1)/n$$

$$V_4 = (V_2 - V_1)/(1 - V_1)$$

$$V_5 = \frac{n}{1+n^2} \left(\frac{1 + V_1}{1 - V_1} \right)$$

$$V_6 = \frac{1}{1+n^2} [V_2 (n \cos \beta - \sin \beta) + nV_1]$$

$$V_7 = \frac{1}{1+n^2} [V_2 (n \sin \beta + \cos \beta) + V_1]$$

$$V_8 = nV_6/(1 - V_1)$$

$$V_9 = nV_7/(1 - V_1)$$

$$V_{10} = \frac{1}{4+n^2} \left[V_2 \left(\frac{n}{2} \sin 2\beta + \cos 2\beta \right) - V_1 \right]$$

$$V_{11} = \frac{1}{4+n^2} \left[V_2 \left(n \cos^2 \beta - \sin 2\beta + \frac{2}{n} \right) - \left(n + \frac{2}{n} \right) V_1 \right]$$

$$V_{12} = \frac{2+n^2}{n(4+n^2)} (1 - V_1) - \frac{n^3 (1+V_1)^2}{(1+n^2)^2 (1-V_1)}$$

$$V_{13} = \frac{n_A}{1+n_A{}^2} (1 + e^{-n} A^\pi)$$

$$V_{14} = \frac{2+n_A{}^2}{n_A (4+n_A{}^2)} (1 - e^{-n}A^\pi)$$

Numerical Examples

Design Problem 31: A cast iron curved member is fixed rigidly at the larger end and is used as a machine frame to support a concentrated end load of $P = 2000$ lbs. Calculate the deflection under the applied load assuming the cross-sectional geometry and the dimensions indicated in Fig. 14-10. Take modulus of elasticity $E = 18 \times 10^6$ psi. Use Lofgren's semigraphical procedure.

Solution: Reference to Fig. 14-10, gives the following moment of

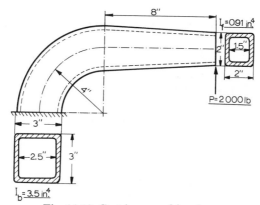

Fig. 14-10. Cast iron machine frame

inertia ratio and length to radius ratio

$$I_b/I_f = 3.5/0.91 = 3.85$$

and

$$L/R = 8/4 = 2$$

From Fig. 14-4, $a/R = 3.13$

and from Fig. 14-5, $\zeta = 17.2$

Hence substituting the numerical data in Eq. (380), gives

$$Y = 17.2 \ \frac{2000 \times 4^3}{18 \times 10^6 \times 3.5}$$

$$Y = 0.035 \text{ in.}$$

Design Problem 32: A single taper steel ring of simple rectangular cross-section is illustrated in Fig. 14-11. Calculate the bending moment at the top of the ring if the ratio of the maximum to the mini-

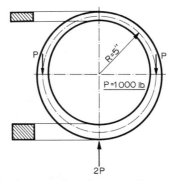

Fig. 14-11. Single taper ring of rectangular cross-section

mum moment of inertia is 4.5. The radius of curvature $R = 5$ in. and the total compressive load is $2P = 2000$ lb.

Solution: From Eq. (382)

$$n = \frac{1}{\pi} \log_e 4.5$$

$$n = 0.48$$

For $\beta = \pi/2$, the bending moment at the top of the ring is given by Eq. (389).

$$M_f = PR \left\{ \frac{(1 - 0.48 V_5)\,[0.48\,V_5\,(V_3 - V_7) + V_{10} - V_6]}{V_{12}} - V_4 + V_9] \right\}$$

From Table 14-1, the required coefficients are

$$V_1 = e^{-0.48\pi} = 0.22$$

$$V_2 = e^{-0.48\pi/2} = 0.47$$

$$V_3 = \frac{(0.47 - 0.22)}{0.48} = 0.52$$

$$V_4 = \frac{(0.47 - 0.22)}{(1 - 0.22)} = 0.32$$

$$V_5 = \frac{0.48\,(1 + 0.22)}{(1 + 0.48^2)\,(1 - 0.22)} = 0.61$$

$$V_6 = \frac{0.48 \times 0.22 - 0.47}{1 + 0.48^2} = -0.30$$

$$V_7 = \frac{0.47 \times 0.48 + 0.22}{1 + 0.48^2} = 0.36$$

$$V_9 = \frac{0.48 \times 0.36}{1 - 0.22} = 0.22$$

$$V_{10} = -\frac{(0.22 + 0.47)}{4 + 0.48^2} = -0.16$$

$$V_{12} = \frac{(2 + 0.48^2)(1 - 0.22)}{0.48(4 + 0.48^2)} - \frac{0.48^3(1 + 0.22)^2}{(1 + 0.48^2)^2(1 - 0.22)} = 0.72$$

Hence substituting the numerical data gives

$$M_f = 5,000 \left\{ \frac{(1 - 0.48 \times 0.61)[0.48 \times 0.61(0.52 - 0.36) - 0.16 + 0.30]}{0.72} \right.$$
$$\left. - 0.23 + 0.16 \right\}$$

$$M_f = 568 \text{ lb-in.}$$

Symbols for Chapter 14

A	Area of cross-section, in.^{2}
A_{max}	Maximum cross-sectional area, in.^{2}
A_{min}	Minimum cross-sectional area, in.^{2}
a	Length, in.
a_x	Reference length, in.
E	Modulus of elasticity, psi
I_x	Moment of inertia for arbitrary section, in.^{4}
I_{max}	Maximum moment of inertia for tapered ring, in.^{4}
I_{min}	Minimum moment of inertia for tapered ring, in.^{4}
I_b	Moment of inertia at built-in end of beam, in.^{4}
I_f	Moment of inertia at free end of beam, in.^{4}
L	Length of straight portion, in.
M	Bending moment, lb-in.
M_f	Fixing moment, lb-in.
M_o	Externally applied bending couple, lb-in.
N	Normal force, lb
n	Taper Factor, Eq. (382)

n_A	Taper factor for ring under internal pressure, Eq. (394)
P	Concentrated load, lb
q	Uniform load, lb/in.
R	Radius to center of gravity, in.
r_o	Radius of gyration of smallest section, in.
s	Length of curved portion; also length along central axis, (Fig. 14-3), in.
U, U_1, U_2	Elastic strain energy, lb-in.
$V_1, V_2 \ldots V_{14}$	Lee's design coefficients
x	Arbitrary distance, in.
Y, Y_1, Y_2	Transverse deflections, in.
θ	Angle at which forces are considered, rad
β	Angle at which load is applied, rad
ζ	Lofgren's deflection factor

CHAPTER FIFTEEN

Design of Circular Arches

Introduction

Continuous increase in complexity of structures and machines calls sometimes for the application of tedious circular arch calculations to advanced design problems. As an example of a specific study consider the derivation of bending moment and deflection equations for circular arches subjected to central loads in the plane of curvature (Refs. 22 & 34). As in the previous cases involving curved members, it is here convenient to use the concept of elastic strain energy and the Castigliano equations. This procedure applies equally well to statically determinate and indeterminate structures. The analysis pertains to arches of uniform cross-section and large radius of curvature in comparison with radial thickness. Strain energy due to bending alone is taken into account and it is assumed that the elastic strain produces no severe deformation under load.

Simply Supported Arch

Consider first a statically determinate case, illustrated in Fig. 15-1. Since there is no constraint of the supports, the reactions are entirely defined by a simple equation of statics and bending moment is found directly in terms of circular geometry and external loading. The behavior of the arch would also be exactly the same if one of the supports were hinged and the other mounted on rollers so that it could move freely in the horizontal direction. It is also assumed that there are no

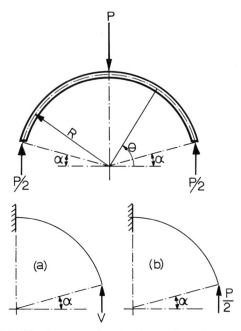

Fig. 15-1. Simply supported circular arch under central load

friction forces, between the supporting surfaces and the ends of the arch, which would oppose horizontal movement. In general such an assumption is justified provided the structure is fairly rigid and the deflection small in comparison with the length of the arch. However in the case of bending very thin metallic strips to large deflections, horizontal constraint due to friction resistance may be of some importance.

The bending moment at a section defined by θ, Fig. 15-1 is

$$M = PR \left(\cos \alpha - \cos \theta \right)/2 \; * \tag{398}$$

Substituting $\theta = \pi/2$ in Eq. (398) gives the maximum bending moment for stress calculations

$$M_{\max} = \frac{PR}{2} \cos \alpha \tag{399}$$

Deflection under load P, is given by the following equation:

$$Y = \frac{2R}{EI} \int_{\alpha}^{\pi/2} M \frac{\partial M}{\partial P} \, d\theta \tag{400}$$

* For meaning of symbols and dimensional units involved for this and other equations in this chapter see material at end of chapter.

where

$$\frac{\partial M}{\partial P} = \frac{R}{2} \, (\cos \alpha - \cos \theta) \tag{401}$$

Hence, substituting Eqs. (398) and (401) into Eq. (400), and integrating, gives

$$Y = \frac{PR^3}{EI} \, G_1 \tag{402}$$

where

$$G_1 = 0.125 \, [(\pi - 2\alpha) \, (1 + 2 \cos^2 \alpha) - 8 \cos \alpha + 3 \sin 2\alpha]$$

Deflection coefficient G_1 is plotted in Fig. 15-2 as a function of the complementary arch angle α. When $\alpha = 0$, $G_1 = 0.1781$ and equation (402) reduces to a standard deflection formula for a semicircular arch. The corresponding maximum bending moment becomes $M_{max} = PR/2$.

Interpretation of Strain Energy

According to the theorem of Castigliano the elastic strain energy of the entire structure is partially differentiated with respect to a selected force, in order to find the displacement of the structure at the point and in the direction of the selected force. Equation (400) is twice the integral between α and $\pi/2$, representing the entire arch shown in Fig. 15-1, while the partial derivative given by Eq. (401) refers to the mid point of the arch in agreement with the basic theorem. Because of symmetry one half of the arch can be analyzed to yield identical results, provided the necessary equations are set up and interpreted correctly. Assume first that one half of the arch, (a) in Fig. 15-1 is acted upon by a force V, equivalent to $P/2$. Hence, we have

$$M = VR \, (\cos \alpha - \cos \theta)$$

$$\partial M / \partial V = R \, (\cos \alpha - \cos \theta)$$

and

$$Y = \frac{VR^3}{EI} \int_\alpha^{\pi/2} (\cos \alpha - \cos \theta)^2 \, d\theta = \frac{2VR^3 G_1}{EI}$$

Since $V = P/2$, the above expression reduces to Eq. (402). If, on the other hand, the cantilever shown at (b) in Fig. 15-1 is analyzed instead, (Eq. (398) remains unaltered, but here we take the partial derivative with respect to $P/2$ and not P as before. This gives

$$\partial M / \partial (P/2) = R \, (\cos \alpha - \cos \theta)$$

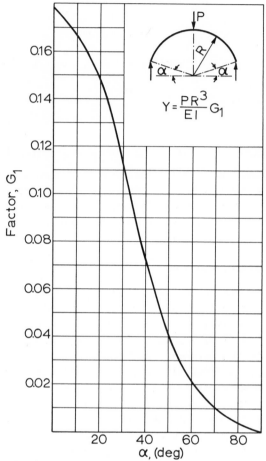

Fig. 15-2. Factor for finding maximum deflection under central load in a simply supported arch

and

$$Y = \frac{PR^3}{EI} \int_a^{\pi/2} \frac{(\cos \alpha - \cos \theta)^2}{2} \, d\theta = \frac{PR^3 G_1}{EI}$$

The above discussion indicates that in dealing with reactions, representing usually statically dependent quantities, it is advisable to employ a different symbol in order to be consistent in setting up the equations according to the theorem of Castigliano. This usually occurs when the original reaction, such as in this case $P/2$, is subsequently treated as an independent force V under which the displacement is sought.

Pinjointed Arch

In considering a circular arch with hinged supports carrying a central vertical load P, the vertical reactions V are equal due to symmetry. The conditions of static equilibrium suggest also that the horizontal reactions H_P must be equal and opposite in direction, but their values cannot be determined from statics. For any point on the arch defined by θ, Fig. 15-3, the bending moment is

$$M = VR(\cos\alpha - \cos\theta) - H_P R(\sin\theta - \sin\alpha) \qquad (403)$$

Here $V = P/2$

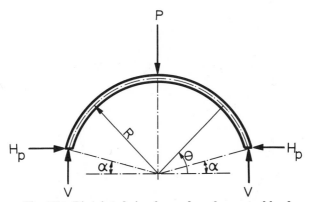

Fig. 15-3. Pinjointed circular arch under central load

Inasmuch as the pinjointed arch is statically indeterminate, H_P is considered a redundant reaction. The value of this redundant quantity can be derived with the aid of the method of least work which is equivalent to the application of Castigliano's second theorem. It is recalled that this theorem states that the deformation of a structure under any load is such that the work of deformation is a minimum. Since in this case the displacement of the arch support is assumed to be zero due to the rigid constraint, the theorem of least work yields

$$\int_a^{\pi/2} M \frac{\partial M}{\partial H_P} d\theta = 0 \qquad (404)$$

Hence, using Eq. (403) gives

$$\partial M/\partial H_P = -R(\sin\theta - \sin\alpha) \qquad (405)$$

Introducing Eqs. (403) and (405) into Eq. (404), integrating and solving for H_P, gives

$$H_P = PA/2B \qquad (406)$$

where

$$A = 4 \sin \alpha + 3 \cos 2\alpha - (\pi - 2\alpha) \sin 2\alpha - 1$$
$$B = (\pi - 2\alpha)(1 + 2 \sin^2 \alpha) - 3 \sin 2\alpha$$

Eq. (406) can be simplified as follows:

$$H_P = PG_2 \qquad (407)$$

When $\alpha = 0$, Eq. (407) gives a standard book value for this case, $H_P = 0.318P$. Thrust coefficient G_2 is given in Fig. 15-4 as a function of the complementary arch angle α. The curve indicates that the

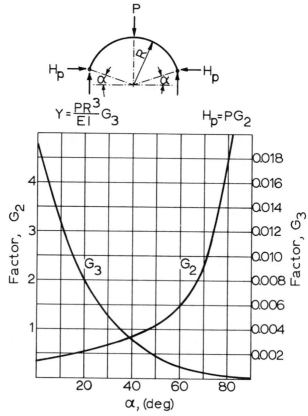

Fig. 15-4. Thrust and deflection factors for pinjointed circular arch under central load

theoretical value of the horizontal thrust increases at a very high rate if there is no yield at the supports and no shortening of the arch due to direct compression. In actual fact the horizontal thrust is always lower than the calculated one, but the obvious implication remains that flat arches are generally subject to appreciable horizontal forces if the supports are kept apart at a fixed distance. For instance, a change in temperature may produce high stresses in a structure with such a horizontal constraint.

The maximum bending moment for the calculation of stresses under load P is obtained now from Eq. (403) at $\theta = \pi/2$, and Eq. (407). This yields

$$M = PR \left[\frac{\cos \alpha}{2} - (1 - \sin \alpha) \ G_2 \right] \tag{408}$$

For $\alpha = 0$, the above moment reduces to $M = 0.182 \ PR$. This value is considerably smaller than that obtained from Eq. (398). In order to find the deflection under load P, consider one half of the arch as an arched cantilever subjected to vertical force V and horizontal thrust H_P, acting as two statically independent forces. This assumption is justified since the external work done by H_P in line of V is zero. The general expression for the deflection due to vertical load follows from the Castigliano equation, Eq. (51a) in which symbol V is substituted for P. From Eq. (403), the partial derivative is

$$\partial M / \partial V = R \ (\cos \alpha - \cos \theta) \tag{409}$$

Hence, combining Eqs. (403), (51a) and (409), integrating and rearranging the terms gives

$$Y = \frac{PR^3}{EI} \ [G_1 - 0.125 \ (A^2/B)] \tag{410}$$

Rewriting this equation yields

$$Y = \frac{PR^3}{EI} \ G_3 \tag{411}$$

G_3 becomes a function of the complementary arch angle α. This function is illustrated in Fig. 15-4. For $\alpha = 0$, Eq. (411) reduces to a standard formula for a semicircular arch with horizontal constraint

$$Y = 0.0189 \ PR^3/EI$$

Built-in Arch

If a circular arch with built-in ends, Fig. 15-5 is submitted to the action of a concentrated vertical load P, at the central point, the

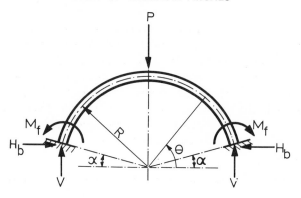

Fig. 15-5. Built-in circular arch under central load

problem of stresses and deflections can be analyzed in a manner analogous to the one just described. Since this arch represents a doubly-redundant structure, the horizontal thrust H_b, and the fixing couple M_f, must be known before the bending moments and deflections can be calculated. The usual procedure in this case is to regard the unknown redundants as loads acting on the primary structure, which due to the symmetry can be subdivided into cantilever arcuate beams fixed at the point of application of the external load. The constraining effects of the arch support are then simulated by the joint action of H_b, M_f and $V = P/2$. The bending moment at any section defined by θ is

$$M = VR \, (\cos \alpha - \cos \theta) - H_b R \, (\sin \theta - \sin \alpha) - M_f$$

$$(412)$$

Since horizontal thrust H_b is assumed not to produce any displacement in its own line of action, the mathematical conditions to be satisfied are the same as those expressed by Eqs. (404) and (405). Therefore utilizing these equations together with Eq. (412) and integrating gives

$$2H_b RB - PRA - 8M_f F = 0 \qquad (413)$$

The other boundary condition involves zero slope at the supports. In mathematical terms this becomes

$$\int_{\alpha}^{\pi/2} M \frac{\partial M}{\partial M_f} \, d\theta = 0 \qquad (414)$$

The partial derivative from Eq. (412) is $\partial M / \partial M_f = -1$.

Hence employing Eqs. (412) and (414) gives

$$2H_b RF - PRS - (\pi - 2\alpha) \, M_f = 0 \qquad (415)$$

Solution of simultaneous Eqs. (413) and (415) yields the following expressions for the redundant reactions:

$$M_f = \frac{PR\,(FA - BS)}{(\pi - 2\alpha)\,B - 8F^2} \tag{416}$$

$$H_b = \frac{P\,[(\pi - 2\alpha)\,A - 8FS]}{2\,(\pi - 2\alpha)\,B - 16F^2} \tag{417}$$

Here

$$F = 0.5\,(\pi - 2\alpha)\,\sin\alpha - \cos\alpha$$

$$S = 1 - 0.5\,(\pi - 2\alpha)\,\cos\alpha - \sin\alpha$$

Coefficients A and B are identical with those employed in Eqs. (406) and (410).

The derivation of the design formula for central deflection can again be accomplished by means of Castigliano equation to give the following result:

$$Y = \frac{PR^3}{EI}\left\{ G_1 - \frac{[4BS^2 - 8AFS + 0.5\,(\pi - 2\alpha)\,A^2]}{4[(\pi - 2\alpha)\,B - 8F^2]} \right\} \tag{418}$$

Finally Eqs. (416), (417) and (418) can be simplified by introducing symbols for the combined trigonometric functions as follows:

$$M_f = PRG_4 \tag{419}$$

$$H_b = PG_5 \tag{420}$$

$$Y = \frac{PR^3}{EI}\,G_6 \tag{421}$$

Coefficients G_4, G_5 and G_6 are illustrated in Fig. 15-6. When $\alpha = 0$, the above equations reduce to standard formulas for a semicircular built-in arch under a central concentrated load

$$M_f = -0.1106\,PR$$

$$H_b = 0.4591\,P$$

$$Y = 0.0117\,PR^3/EI$$

For these conditions the bending moment under load P and $\alpha = 0$, is $M = 0.15\,PR$, indicating compression on the upper surface of the arch in agreement with the original sign convention. The general equation for bending moment under load P is

$$M = PR\,[0.5\cos\alpha - G_5\,(1 - \sin\alpha) - G_4] \tag{422}$$

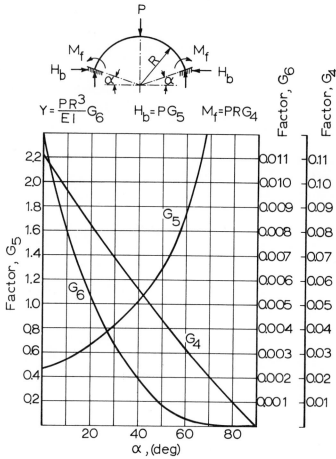

Fig. 15-6. Built-in circular arch: factors for calculating horizontal thrust, fixing couple and vertical deflection

Pinjointed Arch under Uniform Loading

The preceding analysis of circular arches sometimes includes the case of arched elements subjected to inertia loading. As an example consider a pin-jointed semi-circular arch shown in Fig. 15-7. The analysis shows that in this case the horizontal thrust is equal to $H = qR/2$, where q denotes weight of arch per inch of circumference. The bending moment at any section defined by θ is

$$M = 0.5 \, qR^2 \, (\pi - \pi \cos \theta - 3 \sin \theta + 2\theta \cos \theta) \qquad (423)$$

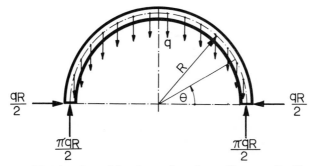

Fig. 15-7. Pinjointed semicircular arch under uniform vertical loading

The maximum deflection is found at $\theta = \pi/2$,

$$Y = 0.0135 \frac{qR^4}{EI} \qquad (424)$$

Eq. (423) indicates that the bending moment undergoes the change in sign at about $\theta = 80$ deg. Also for $\theta = \pi/2$, the bending moment is $M = 0.071\ qR^2$. Positive sign indicates that at $\theta = \pi/2$, upper surface of the arch is in compression.

Symbols for Chapter 15

A	Auxiliary function
B	Auxiliary function
E	Modulus of elasticity, psi
F	Auxiliary function
$G_1, G_2 \ldots G_6$	Force and deflection factors
H_P	Horizontal thrust in pin-jointed arch, lb
H_b	Horizontal thrust in built-in arch, lb
I	Moment of inertia, in.⁴
M	Bending moment, lb-in.
M_f	Fixing moment, lb-in.
$M_{\max}$	Maximum bending moment, lb-in.
P	Vertical load, lb

q	Uniform load, lb/in.
R	Mean radius of curvature, in.
S	Auxiliary function
V	Vertical reaction, lb
Y	Vertical deflection, in.
α	Complementary arch angle, rad
θ	Angle at which forces are considered, rad

CHAPTER SIXTEEN

Special Problems in Curved Member Design

Introduction

The problems of curved member design and analysis discussed so far have been selected largely on the basis of their relative simplicity and applicability to machine construction. In the majority of cases their physical nature could be simulated with the aid of the elementary mathematical models and the amount of computational work, associated with the solutions, was generally moderate.

It is impossible, in the volume of this nature and size, to adequately cover many more complex theoretical problems or even to call attention to all the work of various investigators in the field. This chapter is intended merely as a brief illustration of certain specialized problems which may require either more advanced theoretical analysis or more time consuming numerical procedures. When viewed from the point of view of economy the latter requirement is often difficult to fulfill, since the price to be paid for a more correct physical insight into the core of a problem often becomes prohibitive. However, with the advent of electronic computer capability, more complex analytical models for predicting the behavior of machines are being gradually extended to cover the design situations once of academic interest only.

Ring on Elastic Support

In some technological areas out-of-plane bending of a relatively stiff ring on elastic foundation may be of interest. Usually, the most common cases include, a concentrated force normal to the plane of curvature, externally applied bending couple about a radial axis, and a concentrated twisting moment acting normal to the plane of the ring. The following working formulas for the bending and twisting moments at any point of ring circumference are due to Sadin, Ungar and Shaffer (Ref. 36).

When the ring is much stiffer than the elastic foundation on which it rests, the reactions of the foundation may be obtained from the conditions of equilibrium so that the usual complication of statical indeterminacy can be avoided. The customary assumptions include constant cross-section, small deflections compared with the dimensions of the cross-section, small cross-section in relation to the mean radius of the ring, and applicability of Hooke's law and the principle of superposition.

A stiff ring resting on an elastic foundation and carrying a single external load P is shown in Fig. 16-1. The reaction of the foundation may be assumed to consist of two components. One is uniformly distributed load $q = -P/2\pi R$. The other resists the moment PR and may be distributed according to a cosine function $q = q_m \cos \theta$, that gives

$$\int_0^{2\pi} (q_m \cos \theta) \ (R \cos \theta) \ R \ d\theta = PR \ *$$

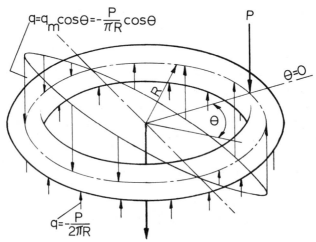

Fig. 16-1. Ring on elastic foundation under a concentrated load

* For meaning of symbols and dimensional units involved for this and other equations in this chapter see material at end of chapter.

from which

$$q_m = -P/\pi R$$

Using this type of loading Sadin, Ungar and Shaffer (Ref. 36) derived the following equations for the bending and twisting moments at any point of the ring.

$$M = -\frac{PR}{4\pi} [2\ (\theta - \pi) \sin\theta + \cos\theta + 2] \qquad (425)$$

and

$$T = -\frac{PR}{4\pi} [3 \sin\theta + 2\ (\theta - \pi)\ (1 - \cos\theta)] \qquad (426)$$

When $\theta = 0$, the bending moment given by Eq. (425) attains a maximum value, $M = -3PR/4\pi$, and zero twisting moment. The absolute values of the moments are symmetric with respect to the diameter of the ring passing through $\theta = 0$.

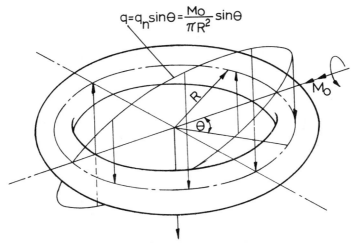

$$q = q_n \sin\theta = \frac{M_o}{\pi R^2} \sin\theta$$

Fig. 16-2. Ring on elastic foundation under a single external couple

When concentrated bending moment is applied about a radial axis passing through $\theta = 0$, Fig. 16-2, a sinusoidally distributed reaction $q = q_n \sin\theta$, may be assumed. Hence

$$\int_0^{2\pi} (q_n \sin\theta)\ (R \sin\theta)\ R\ d\theta = M_o$$

from which

$$q_n = M_o/\pi R^2$$

The corresponding moments at any point of the ring are shown to be (Ref. 36),

$$M = \frac{M_o}{4\pi} [2 (\pi - \theta) \cos \theta - \sin \theta] \qquad (427)$$

and

$$T = \frac{M_o}{4\pi} [2 (\pi - \theta) \sin \theta - \cos \theta - 2] \qquad (428)$$

Eqs. (427) and (428) indicate that the maximum bending and twisting moments occur at $\theta = 0$.

For an externally applied twisting couple T_o as shown in Fig. 16-3, the supporting foundation must exert an equal and opposite moment

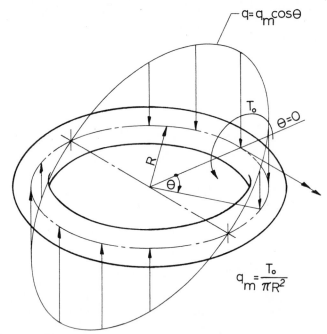

Fig. 16-3. Ring on elastic foundation under a single twisting moment

on the ring which may be assumed to vary according to a cosine function $q = q_m \cos \theta$, which gives, as before, $q_m = T_o/\pi R^2$. The moment equations for this case are

$$M = \frac{T_o}{4\pi} \left[2 (\theta - \pi) \sin \theta + \left(\frac{1 - 3\lambda}{1 + \lambda} \right) \cos \theta \right] \qquad (429)$$

and

$$T = \frac{T_o}{4\pi}\left[\left(\frac{3-\lambda}{1+\lambda}\right)\sin\theta - 2\,(\theta-\pi)\cos\theta\right] \qquad (430)$$

The distribution of bending moments given by the above equations depends not only on θ, but also on the parameter λ, which will be somewhat different for various materials and cross-sectional geometry. It is recalled here that λ, denotes the ratio of flexural to torsional rigidity.

Design of Pipe-bend under Uniform Loading

In designing pipework for stresses and deformation due to thermal expansion or movement of supports various design equations and methods are available in the literature (Refs. 37 and 38). Such equations are usually based on the modifications of elastic arch theory where either the forces or displacements at the supports are specified.

When a piping system is subjected to uniformly distributed loading, such as that caused by flow inertia or other dynamic effects, the equations for parametric studies of statically indeterminate reactions at the supports are not readily available. Consider for instance a simple pipe bend buit-in at the supports and uniformly loaded in x and y directions as shown in Fig. 16-4. Let us assume that the constraint at A can be simulated by the combined action of normal force N, shear force Q and fixing couple M_f. Since these reactions are statically indeterminate the second theorem of Castigliano can be

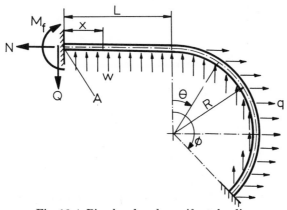

Fig. 16-4. Pipe bend under uniform loading

used to select such values of the reactions for which the vertical deflection, horizontal deflection and slope at A are equal to zero. Mathematically this condition is stated as follows:

$$\frac{\partial U}{\partial N} = \frac{\partial U}{\partial Q} = \frac{\partial U}{\partial M_f} = 0$$

Since the pipe bend in this case is relatively flexible, U denotes the elastic strain energy due to bending.

For an auxiliary angle ε, smaller than θ, measured from axis $\theta = 0$, the elementary bending moment due to the horizontal unit load q, is

$$dM_q = qR^2 (\cos \varepsilon - \cos \theta)\, d\varepsilon$$

Integrating the above expression between 0 and θ gives

$$M_q = qR^2 (\sin \theta - \theta \cos \theta) \tag{431}$$

The elementary bending moment due to the vertical loading w is

$$dM_w = wR^2 (\sin \theta - \sin \varepsilon)\, d\varepsilon$$

On integration this gives

$$M_w = wR^2 (\theta \sin \theta + \cos \theta - 1) \tag{432}$$

If the total bending moments for the straight and curved portions of the pipe are denoted by M_1 and M_2, respectively, utilizing notation given in Fig. 16-4 and Eqs. (431) and (432) yields

$$M_1 = \frac{wx^2}{2} - Qx - M_f \tag{433}$$

$$M_2 = wR^2 (\theta \sin \theta + \cos \theta - 1) + wL \left(\frac{L}{2} + R \sin \theta \right)$$
$$+ qR^2 (\sin \theta - \theta \cos \theta) - Q (L + R \sin \theta)$$
$$- NR (1 - \cos \theta) - M_f \tag{434}$$

The boundary conditons at A are

$$\int_0^L M_1 \frac{\partial M_1}{\partial M_f} dx + \int_0^\phi M_2 \frac{\partial M_2}{\partial M_f} R\, d\theta = 0 \tag{435}$$

$$\int_0^\phi M_2 \frac{\partial M_2}{\partial N} d\theta = 0 \tag{436}$$

$$\int_0^L M_1 \frac{\partial M_1}{\partial Q} dx + \int_0^\phi M_2 \frac{\partial M_2}{\partial Q} R\, d\theta = 0 \tag{437}$$

The relevant partial derivatives follow from Eqs. (433) and (434)

$$\frac{\partial M_1}{\partial M_f} = \frac{\partial M_2}{\partial M_f} = -1$$

$$\frac{\partial M_2}{\partial N} = -R\,(1 - \cos\theta)$$

$$\frac{\partial M_1}{\partial Q} = -x$$

$$\frac{\partial M_2}{\partial Q} = -\,(L + R\sin\theta)$$

Substituting the above derivatives and Eqs. (433) and (434) into the boundary expressions, Eqs. (435), (436) and (437) yields the following simultaneous equations for calculating statically indeterminate quantities, (Ref. 39):

$$2M_f\,(k + \phi) + QR\,(k^2 + 2k\phi - 2\cos\phi + 2) + 2NR\,(\phi - \sin\phi)$$
$$+ 2qR^2\,(2\cos\phi + \phi\sin\phi - 2) + wR^2\,(2\phi\cos\phi + 2\phi - 4\sin\phi$$
$$- k^2\phi + 2k\cos\phi - 2k) = 0 \qquad (438)$$

$$12M_f\,(k^2 + 2k\phi - 2\cos\phi + 2) + 2QR\,(4k^3 + 12k^2\phi - 24k\phi$$
$$+ 24k + 6\phi - 3\sin 2\phi) - 12NR\,(2k\phi + 2k\sin\phi + 2\cos\phi$$
$$+ \sin^2\phi - 2) - 3qR^2\,(4\phi - 3\sin 2\phi + 2\phi\cos 2\phi)$$
$$+ 3wR^2\,(12k^2\cos\phi - 12k^2 - 2\phi^2 + 2\phi\sin 2\phi + 3\cos 2\phi + 5$$
$$- 8\cos\phi - 16k\sin\phi + 8k\cos\phi + 4k\phi + 2k\sin 2\phi) = 0 \qquad (439)$$

$$8M_f\,(\phi - \sin\phi) + 4QR\,(2k\phi - 2k\sin\phi - 2\cos\phi + 2 - \sin^2\phi)$$
$$+ 2NR\,(6\phi - 8\sin\phi + \sin 2\phi) + qR^2\,(16\cos\phi + 8\phi\sin\phi$$
$$- 2\phi^2 - 2\phi\sin 2\phi - 3\cos 2\phi - 13) + wR^2\,(8\phi\cos\phi + 3\sin 2\phi$$
$$- 2\phi\cos 2\phi - 24\sin\phi + 12\phi - 4k^2\phi + 4k^2\sin\phi + 8k\cos\phi$$
$$+ 4k\sin^2\phi - 8k) = 0 \qquad (440)$$

Solution of the above equations is best obtained for a specific value of length ratio $k = L/R$, and subtended angle ϕ. When $k = 0$, $w = 0$ and $\phi = \pi$, solution of Eqs. (438) through (440) gives the reactions for a semicircular arch under transverse uniform load.

$$N = \pi q R/2$$

$$Q = \frac{3\pi^2 - 32}{2\,(\pi^2 - 8)}\,qR$$

$$M_f = \frac{\pi\,(10 - \pi^2)}{2\,(\pi^2 - 8)}\,qR^2$$

Substituting $k = 0$, $q = 0$ and $\phi = \pi$, yields on the other hand the design formulas for redundant reactions in a semicircular arch subjected to uniform sideload.

$$N = -wR/2$$
$$Q = \pi wR/2$$
$$M_f = -wR^2/2$$

Buckling Column Spring

When an initially straight and slender member of uniform cross section is subjected to a compressive end load, and when the critical column load is exceeded the deformation begins to increase smoothly at a constant rate. The straight line portion of the load-deflection characteristics develops a relatively low gradient and the column can be utilized in certain retaining mechanisms and low frequency machine elements where small working loads and large deflections are required. Because of large deflections the initially straight column behaves as a curved member and the differential equation applicable to this case, Fig. 16-5, is

$$EI\,(d\psi/ds) = -Px \qquad (441)$$

Here, ψ denotes slope at a point of deflection curve, s is measured along the length of curved member, and Px denotes the bending moment due to the compressive load P. Solution of Eq. (441) involves elliptic integrals which can be simplified to give formulas and charts for the calculation of load-deflection characteristics for a buckling column spring, (Ref. 40). For a slender member of rectangular cross-section the buckling load in the range of large deflections becomes

$$P = \frac{AE\Phi}{m^2} \qquad (442)$$

In Eq. (442), A is cross-sectional area of buckled member, E the modulus of elasticity, Φ the load deflection factor given in Fig. 16-6 and m denotes the ratio of column length L to depth of cross-section h.

It may be noted that load-deflection factor Φ, given in Fig. 16-6, can be approximated by a straight line relationship between the values of Y/L, ranging from 0.05 to 0.40. The experimental analysis and calculations (Ref. 40) indicate that 0.05 and 0.40 may be recom-

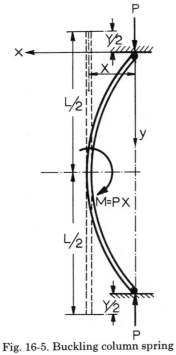

Fig. 16-5. Buckling column spring

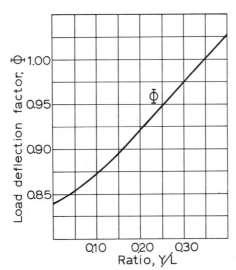

Fig. 16-6. Load-deflection factor for a buckling column spring of uniform
rectangular cross-section

mended as practical design limits for this type of load carrying member.

When the longitudinal deflection Y, is calculated with the aid of formula, Eq. (442) and design curve, Fig. 16-6, the designer may wish to find the corresponding value of lateral displacement X. Since a unique relation exists between the lengthwise and lateral displacements for the load carrying member under consideration, a simple conversion chart is given in Fig. 16-7 for rapid calculations.

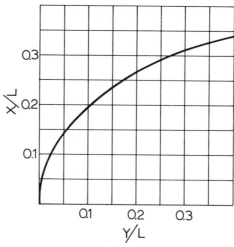

Fig. 16-7. Relation between longitudinal and lateral deflections for a buckling column spring of uniform rectangular cross-section

When a relatively thin compression member is loaded between the pivot supports, appreciable bending stresses may develop, even for a moderate load capacity. Whereas the derivation of the deflection can not be based on the theory of small deflections, an idea of the maximum bending stress involved may be had using the elementary beam analysis. As a first approximation the assumption is made that the maximum bending moment increases in direct proportion to the lateral displacement X, which can be calculated using the above design procedure. The stress is then found by the usual formula $S = M/Z$. Here, simple calculations will indicate that, in certain cases, permanent set of the buckling column spring should be expected for higher ratios of Y/L.

It is well to note also that great care should be exercised in selecting the thickness of stock the tolerance of which is often difficult to maintain in manufacture and measure with sufficient degree of accuracy. This is especially important in designs utilizing thin stock since the compressive end-load P is then directly proportional to the cube of thickness.

In the foregoing method presented for the buckling column spring, behaving as a special curved member, relatively large deflections were admitted which implied that the material must exhibit a high yield point to avoid the development of permanent set. Furthermore, the column member was assumed to be perfectly straight initially and the loads applied concentrically. To clarify the important matter of initial imperfections in the column spring, Masur * derived special formulas for the buckling load and the maximum bending stress, using the principle of virtual work and representing "buckled shape" of the column spring by a circular arc. Expanding the trigonometric functions in Taylor series, in the region of $\theta = 0$, and retaining the lowest terms, Masur's load formula is

$$P = \frac{12EI}{L^2} \frac{(\phi - \phi_o)\,(10 + \phi^2)}{10\phi} \tag{443}$$

In Eq. (443), ϕ and ϕ_o, are one half the central angles of the column bent to a circular arc in its initial and final state, respectively. When $\phi_o = 0$, the result is about 20 percent larger than that obtained from Eq. (442). The difference is due to the choice of the buckling mode. The corresponding Masur's stress formula is

$$S = \frac{E}{m^2}\left[1 - \frac{1}{\sqrt{1 + \dfrac{3YL}{8X_o{}^2}}}\right]\left(1 + \frac{0.6Y}{L}\right) \tag{444}$$

In Eq. (444), X_o denotes the initial amplitude of imperfection. When L/X_o becomes infinitely large, Eq. (444) represents the maximum bending stress for a buckling column spring which is initially perfectly straight. A study of the parameter Sm^2/E as a function of L/X_o and Y/L indicates that the effect of initial imperfections is strongest for very small values of Y/L. Within the range, for which Eq. (442) is recommended, the effect of initial imperfections appears to be relatively small. Practical experience with buckling column spring shows that the best results are obtained when Y/L varies between 0.1 and 0.3, and L/h ratio is about 200, or more.

---

* Written contribution by E. F. Masur, University of Illinois, to the discussion of Ref. 40.

Symbols for Chapter 16

A	Cross-sectional area, in.^{2}
E	Modulus of elasticity, psi
h	Depth of cross-section, in.
I	Moment of inertia, in.^{4}
$k = L/R$	Length to radius ratio
L	Straight length; length of buckling column, in.
M	Bending moment, lb-in.
M_o	Externally applied bending couple, lb-in.
M_q, M_w	Bending moments due to uniform loading, lb-in.
M_1, M_2	Bending moments for straight and curved portions, lb-in.
M_f	Fixing moment, lb-in.
m	Ratio of length to thickness for buckling column
N	Normal force, lb
P	Vertical load, lb
Q	Shearing force, lb
q	Uniform load, lb/in.
q_m, q_n	Maximum unit loads, lb/in.
R	Mean radius of curvature, in.
S	Stress, psi
s	Arc length; length measured along curved member, in.
T	Twisting moment, lb-in.
T_o	Externally applied twisting couple, lb-in.
U	Elastic strain energy, lb-in.
w	Uniform load, lb/in.
X	Horizontal deflection, in.
X_o	Initial amplitude of imperfection, in.

x	Arbitrary distance, in.
Y	Vertical deflection, in.
y	Arbitrary distance, in.
Z	Section modulus, in.^{3}
ε	Auxiliary angle, rad
θ	Angle at which forces are considered, rad
λ	Ratio of flexural to torsional rigidity
ϕ, ϕ_o	Angles subtended by curved member, rad
Φ	Load deflection factor for buckling column
ψ	Slope, rad

References

1. TIMOSHENKO, S., and GOODIER, J. N., *Theory of Elasticity*, Mc-Graw-Hill Book Company, Inc., New York, 1951.
2. GOODENOUGH, G. A., and MOORE, L. E., "Strength of Chain Links" Bulletin 18, Engineering Experimental Station, University of Illinois, Urbana, Illinois, 1907.
3. MORLEY, A., "Bending Stresses in Hooks and Other Curved Beams," *Engineering*, Engineering Ltd., London WC2, England, Vol. 98, 1914.
4. TIMOSHENKO, S., "On the Distribution of Stresses in a Circular Ring Compressed by Two Forces Acting along a Diameter," *Philosophical Magazine*, England, Vol. 44, No. 263, p. 1014, 1922.
5. PIPPARD, A. J. S., and MILLER, C. V., "The Stresses in Links and their Alteration in Length under Load," *Proceedings of the Institution of Mechanical Engineers*, England, 1923.
6. WINSLOW, A. M., and EDMONDS, R. H. G., "Tests and Theory of Curved Beams," Transactions of The American Society of Mechanical Engineers, p. 647, 1926.
7. ANDREWS, E. S., "Elastic Stresses in Structures," Scott, Greenwood, London, England, 1919.
8. LEEMAN, E. R., "Stresses in a Circular Ring," *Engineering*, Engineering Ltd., London, WC2, England, April 13, 1956.
9. HETENYI, M., *Handbook of Experimental Stress Analysis*, John Wiley and Sons, Inc., New York, 1950.
10. SHIGLEY, J. E., *Machine Design*, McGraw-Hill Book Company, Inc., New York, 1956.
11. ROARK, R. J., *Formulas for Stress and Strain*, McGraw-Hill Book Company, Inc., New York, 1954.
12. PETERSON, R. E., *Stress Concentration Design Factors*, John Wiley and Sons, Inc., New York, 1953.

13. BOYD, J. E. and FOLK, S. B., *Strength of Materials*, McGraw-Hill Book Company, Inc., New York, 1950.

14. MARIN, J., *Mechanical Behavior of Engineering Materials*, Prentice-Hall, Inc., Englewood Cliffs, New Jersey, 1962.

15. OBERG, E. and JONES, F. D., *Machinery's Handbook*, Seventeenth Edition, The Industrial Press, New York, 1964.

16. POPOV, E. P., *Mechanics of Materials*, Prentice-Hall, Inc., Englewood Cliffs, New Jersey, 1957.

17. TIMOSHENKO, S., *Strength of Materials*, D. Van Nostrand Company, Inc., Princeton, New Jersey, 1956.

18. BORG, S. F. and GENNARO, J. J., *Advanced Structural Analysis*, D. Van Nostrand Company, Inc., Princeton, New Jersey, 1959.

19. BLEICH, F., *Buckling Strength of Metal Structures*, McGraw-Hill Book Company, Inc., New York, 1952.

20. THE LINCOLN ELECTRIC COMPANY, *Procedure Handbook of Arc Welding Design and Practice*, Eleventh Edition, Cleveland, Ohio, 1957.

21. BLAKE, A., "Arched Cantilever Beams," *Machine Design*, The Penton Publishing Company, Cleveland, Ohio, June 26, 1958.

22. BLAKE, A.,"Rings and Arcuate Beams," *Product Engineering*, McGraw-Hill Publishing Company, New York, January 7, 1963.

23. PALM, J. and THOMAS, K., "Berechnung gekrümmter Biegefedern," VDI-Z, Bd. 101, Nr. 8, 1959.

24. BLAKE, A., "Curved-end Cantilevers," *Machine Design* (Data Sheets), The Penton Publishing Company, Cleveland, Ohio, 1959.

25. BLAKE, A., "Complex Flat Springs," *Product Engineering*, McGraw-Hill Publishing Company, New York, October 2, 1961.

26. FLECKENSTEIN, J. E., "U-Springs – Stress and Deflection Calculations." *ASME Paper*, 60-WA-172, 1960.

27. BIEZENO, C. B. und GRAMMEL, R., *Technische Dynamik*, Springer-Verlag, Berlin/Göttingen/Heidelberg, Germany, 1953.

28. McGUINESS, H. D., "Solution of a Circular Ring Structural Problem," Technical Report 32-178, California Institute of Technology, 1961.

29. BLAKE, A.,"The Calculation of the Deflection of Circular Arches and a Study of the Deflection of Thick Rings," M.Sc. Thesis, London University, London, England, 1955.

30. WILSON, B. J., and QUEREAU, J. F., "A Simple Method of Determining Stress in Curved Flexural Members," Circular 16, Engineering Experimental Station, University of Illinois, Urbana, Illinois, 1927.

31. BLAKE, A.,"Deflection of a Thick Ring in Diametral Compression by Test and by Strength of Materials Theory," ASME Journal of Applied Mechanics, June 1959.
32. LOFGREN, K. E., "Calculating Deflection of Curved Beams," *Machine Design*, The Penton Publishing Company, Cleveland, Ohio, November 1948.
33. LEE, TEH H., "Varying Section Rings," *Product Engineering*, McGraw-Hill Publishing Company, New York, July 6, 1964.
34. BLAKE, A., "Circular Arches," *Machine Design*, The Penton Publishing Company, Cleveland, Ohio, December 25, 1958.
35. TABAKMAN, H. D., and VALENTIJN, H. P., "Distortion of Circular Rings," *Machine Design*, The Penton Publishing Company, Cleveland, Ohio, October 22, 1964.
36. SADIN, S. R., UNGAR, E. E., and SHAFFER, B. W., "Out-of-plane Bending of a Relatively Stiff, Elastically Supported Ring," New York University, 1956.
37. BAUMEISTER, T., and MARKS, L. S., *Mechanical Engineer's Handbook*, Sixth Edition, McGraw-Hill Book Company, Inc., New York, 1958.
38. THE M. W. KELLOG COMPANY, *Design of Piping Systems*, John Wiley and Sons, Inc., New York, 1956.
39. BLAKE, A., "Flexibility and Strength Considerations for Engine Lines," Technical Memorandum, Aerojet-General Corporation, Sacramento, California, 1964.
40. BLAKE, A., "Analysis of Buckling Column Spring with Pivoted Ends and Uniform Rectangular Cross-section," ASME Journal of Engineering for Industry, Paper No. 60-SA-10, 1961.

Index

DATE DUE

HETERICK MEMORIAL LIBRARY
621.815 B636d   onuu
Blake, Alexander/Design of curved member

3 5111 00122 8687